ROUTLEDGE LIBRARY EDITIONS: AGRICULTURE

Volume 6

CHANGING PATTERNS IN ISRAEL AGRICULTURE

CHANGING PATTERNS IN ISRAEL AGRICULTURE

HAIM HALPERIN

Routledge
Taylor & Francis Group

LONDON AND NEW YORK

First published in 1957 by Routledge & Kegan Paul Ltd.

This edition first published in 2020
by Routledge
2 Park Square, Milton Park, Abingdon, Oxon OX14 4RN

and by Routledge
52 Vanderbilt Avenue, New York, NY 10017

Routledge is an imprint of the Taylor & Francis Group, an informa business

British Library Cataloguing in Publication Data
A catalogue record for this book is available from the British Library

ISBN: 978-0-367-24917-5 (Set)
ISBN: 978-0-429-32954-8 (Set) (ebk)
ISBN: 978-0-367-25683-8 (Volume 6) (hbk)
ISBN: 978-0-367-25685-2 (Volume 6) (pbk)
ISBN: 978-0-429-28913-2 (Volume 6) (ebk)

Publisher's Note
The publisher has gone to great lengths to ensure the quality of this reprint but points out that some imperfections in the original copies may be apparent.

Disclaimer
The publisher has made every effort to trace copyright holders and would welcome correspondence from those they have been unable to trace.

CHANGING PATTERNS IN ISRAEL AGRICULTURE

By

HAIM HALPERIN

Routledge and Kegan Paul

LONDON

First published 1957
by Routledge & Kegan Paul Ltd.
Broadway House, Carter Lane, E.C.4
Printed in Great Britain
by Butler & Tanner Ltd.
Frome and London

CONTENTS

CONTENTS

MAP AND FIGURES

PREFACE

For seventy-five generations the Jewish people waged an unremitting struggle for national independence and liberty, but it was vouchsafed only to the present generation to consummate the ancient ideal and to establish the State of Israel. The task of realization has been arduous. The very process of expanding the population twofold in a compressed period of three and a half to four years inevitably produced currents of social and economic change whose reverberations will be felt for many years to come. The many and varied tribes of Israel, the product of as many cultures and civilizations, are uniting to form a single nation. The instruments of national sovereignty are being developed: a national army, an administration, a national economy. A complex of agriculture, industry, public services is emerging, a distinctive culture is in the making.

Agriculture, beyond question, is one of the most important areas of society. The significance cannot be measured solely by its contribution towards satisfying the most elementary of human needs—food; it must also be gauged by its place in the political life of the country. In Israel, it assumes a peculiar significance because the whole process of Jewish return to recognized nationhood was conceived as a return to the land. A nation without peasants rooted in the soil it was held was an inconceivable abstraction. This tenet shaped the character and the course of the national movement and even before Israel became a sovereign entity the roots in the soil had already been struck.

It was against this background that the immigrants had to adjust themselves to their new conditions after independence was achieved. But the new dimensions of mass-immigration and mass-colonization generated new trends in Israel agricultural settlement involving structural, social and economic changes in farming and in rural life generally.

The present study seeks to analyse the problems raised by these new forces and conditions. It discusses the impact of Israel's achievement of political sovereignty upon its agricultural economy in the comparatively short space of six years; it examines the agricultural problems that have arisen as functions of the natural factors of production—land, water, climate, etc.; it endeavours to assess new and better possibilities of farming.

AGRICULTURAL GEOGRAPHY
OF ISRAEL

ISRAEL, though one of the smallest countries in the world, is remarkable for its topographical and climatic diversity. At one stage
Israel's settlement planners, in their desire to secure areas more
or less homogeneous, contemplated sub-dividing the country into
thirty-four planning regions. (In Israel it is not unusual to find
several types of soil within the area of a single settlement.) Agricultural planning at present, however, is being conducted within the
framework of twelve districts (comprising twenty-seven sub-districts)
into which the country has been partitioned. But even this arrangement has proved unsatisfactory, as in some districts there are areas
which though conjacent, are no less disparate than highland and
valley.

It is our opinion that the most suitable division for agricultural
purposes would be into at least eighteen districts.

The area of Israel is 20,800 square kilometres (8,030 square miles).[1]
Within this small area differences in altitude range from 1,208 metres
(3,963 feet) above sea-level at Mount Atzmon in Galilee, to 392
metres (1,286 feet) below sea-level at the desolate shores of the Dead
Sea and 209 metres (686 feet) below sea-level along the closely
settled banks of Lake Kinneret.[2]

Another indication of the great disparity in natural conditions is
provided by the rainfall, which ranges from a maximum of 1,100
millimetres (44 inches) on Mount Atzmon and 700–800 millimetres
(28–32 inches) in Central Upper Galilee to 25 millimetres (1 inch)
and even less in the desert region of the Negev. In between are the

---

[1] *Division of Land Survey*, Ministry of Labour, Israel Government.
[2] *Statistical Abstract of Israel*, 1953/54, Table 1.

highlands with a rainfall of 600–800 millimetres (24–32 inches), the Coastal Plain with 370–550 millimetres (15–22 inches) and the Southern District within which again there are also wide differences, such as between Asluj—120 millimetres (5 inches)—and Beersheba—220 millimetres (9 inches), although they are only 28 kilometres apart.[1]

The disparity in respect of mean relative humidity—measured over a period of ten years—is also remarkable, as is indicated by the figure for the Jordan Valley—52%—and that for Tel-Aviv—71%.[1]

From the geologist's point of view Israel is a 'young' country. Its important strata represent the last 8% of geological time. The basic rock lies in horizontal layers at a depth of 900 metres (3,000 feet approximately.) Mostly, however, the superjacent layers have shifted vertically. The rifts occur from east to west and from south-west to north-west. The greatest uplift occurs in a north–south direction, from the Syrian Plateau southwards by way of the Red Sea to East Africa. Here, along parallel rifts, the central section has subsided to form a strip of varying width, which in Israel has assumed the form of a deep and broad valley, the Jordan Depression. This rift in Israel drops to the lowest altitude on the earth's surface, some 392 metres (1,286 feet) below sea-level.

It is generally held that in later geological time there existed no outlet for the water of the Jordan Depression other than evaporation, with the result that a vast cumulation of brine, the Dead Sea, was created.

During the same period as the formation of the Jordan Depression a similar process took place in the west of Israel. Layers of soil subsided step-fashion. The outermost western strip slumped beneath the sea. There were also layers which subsided to create cross-rifts. One of these depressions is the Vale of Jezreel, another separates the Judean highlands from the Negev.

Extensive volcanic areas in Galilee were produced by immense earthquakes which formed the existing topography.

The movements shaping Israel's topography have not ceased. The country belongs to the area of seismic disturbances and as recently as 1927 a serious earthquake occurred.

In addition to geological factors, water, winds, animals and men have all exerted an influence in shaping the physical features of the country, and have co-operated in various periods in destroying the upper layer of soil.

The waves of the Mediterranean created sand-dunes, the winds of the Negev deposited the layer of loess, which was carried here—and

---

[1] *Statistical Abstract of Israel*, 1953/54, Table 2 and others.

is still being carried—from afar. Floodwaters and winds eroded the fertile soil covering the mountain slopes. Men and goats destroyed the grasses and the trees, which had previously checked the inroads of erosion, thereby increasing the havoc.

Within a comparatively short distance striking differences in the nature of countryside are perceptible, no less indeed than might distinguish diverse countries and climates. Within a matter of hours the traveller passes from the desert with its mirages to verdant citrus groves, from bald, denuded mountains to fertile valleys, watered by sprinklers because the farmers do not wish to rely upon the caprices of the climate.

The agricultural problems which this heterogeneity in nature has begotten are many and complex and each successive generation in this country has sought an adequate solution with the means and the techniques at its disposal.

CLIMATIC CONDITIONS

Climatically, Israel can be divided into four clear demarcated zones:

(1) The Coastal Plain, which is warm and humid in summer and mild in winter, with westerly winds blowing from the Mediterranean. Daily and annual fluctuations in temperature are not extreme. Rainfall is adequate, the average over forty years being 611 millimetres (25 inches) on 51 days at the northern station of the zone, and 550 millimetres (22 inches) on 60 days at the southern station.

(2) The hill country is drier and cooler than the Coastal Plain in summer and cold in winter. Daily and yearly temperature fluctuations are sharper. The mean annual rainfall, based upon observations over forty years, is 638 millimetres·(26 inches) on 52 days at the Nazareth station, and 583 millimetres (23 inches) on 59 days in Jerusalem.

(3) The Jordan Valley is warm and dry, without moderating westerly winds in the summer. In winter temperature and humidity are medium. There is little rainfall—305 millimetres (12 inches) on 38 days in the northern section, and 152 millimetres (6 inches) on 31 days in the southern section, based on observations over forty years.

(4) The Negev is a semi-arid zone, which becomes arid south of Beersheba down to Elath. It is hot and dry in summer, cold and dry in winter. There are sharp fluctuations in both daily and annual temperatures and very little rainfall. At the Beersheba station the annual average over forty years is 220 millimetres (9 inches) over 23 days.

Israel, lying between latitudes 29° 30' and 33° 15' North, is

3

TABLE 1

Fluctuations of temperature in Israel (C)

	Annual average	Mean maximum for hottest month	Mean minimum for coldest month
Haifa	21·0	29·4	9
Tel-Aviv	20·5	32·2	9
Jerusalem	17·8	29·4	4
Beersheba	19·4	33·3	5

situated in the sub-tropical zone. In respect of rainfall, however, two-thirds of its area, with an annual precipitation of less than 350 millimetres (14 inches), is in the arid zone and the remainder in the semi-arid zone.

The high-lying areas in Judea and Galilee with a heavy annual rainfall ranging between 750 and 1,000 millimetres (30–40 inches) belong to the humid region, but the areas bordering upon these highlands, which are more extensive and have a mean annual rainfall of 550–750 millimetres (22–30 inches), can be considered to belong to the semi-humid region. Most of Israel, which belongs to the Mediterranean littoral, has a typically Mediterranean climate and countryside emphasized by conditions of semi-aridity, few valleys, lack of drainage, endemic malaria, soil suitable for orchards rather than field crops and with few forests or fishing grounds.

Even on the loftier mountains (which, indeed, are not high compared to mountains in the other countries), where the amount of rainfall equals that of the semi-humid zone, a variety of natural features, such as the number of days of rain, the porousness of the soil, high seasonal and annual temperatures are more typical of the semi-arid zone.

Despite the heavy precipitation in the highlands—high not only relative to other districts of the country but in comparison with London with its annual rainfall of 615 millimetres (25 inches) or Paris with 575 millimetres (23 inches), as against Jerusalem with 660 millimetres[1] (26 inches)—the mountainous regions of Israel are regarded as inadequately watered. The reasons are as follows:

(a) The distribution of rainfall is not favourable. A large volume of rain falls in a comparatively short season. The year comprises two sharply divided seasons—a dry summer lasting seven months, during which a permanent anti-cyclone prevents any rainfall whatever, and the cyclonic season in winter lasting five months, in which rainfall is not evenly distributed, several weeks often elapsing between

[1] Robert R. Nathan, Oscar Gass and Daniel Creamer, *Palestine Problem and Promise*, American Council on Public Affairs, 1946, p. 108.

one spell of rain and the next. The average number of days of rain in the districts with heavy rainfall is approximately sixty.

(b) The rainwater, for the most part, is not absorbed in the soil because of the bareness of the hillsides, which in turn is the result of many generations of destructive farming methods and particularly over-grazing by herds of goats.

TOPOGRAPHICAL REGIONS

The country can be divided into six main topographical regions:

(1) The Coastal Plain, stretching from north of the Egyptian border, along the Lowland, Sharon and Samarian plains to the Carmel coast near Haifa, and from there through the Vale of Zebulon and Acre to Rosh Hanikra on the Lebanese border.

(2) The Judean Mountains and Samaria.

(3) The Jordan Depression.

(4) The Negev.

(5) The Vale of Jezreel (Esdraelon).

(6) The Galilean Mountains.

A clear line of demarcation is drawn by the Jordan Depression which runs from north to south and divides the Judean Mountains from the Transjordanian Plateau.

West of the Judean Mountains lies the Coastal Plain. North of these mountains and Samaria runs the Vale of Esdraelon, separating the central massif from the Galilean Mountains.

The Negev lies to the south of the Judean Mountains.

The valleys are inter-connected and in effect constitute an unbroken stretch of country running from north to south. The mountains, too, constitute great massifs broken only twice—in the north by the Vale of Jezreel and in the south by the Judean Plain down to the Negev Highlands. On both sides of the mountains there are valleys and plains.

TYPES OF SOIL

The principal rocks from which the soils of Israel have developed are chalk, dolomite, granite, basalt, sandstone, etc.

These soils can be classified on the basis of their origin as follows:

(1) *Alluvial clay soils*—in areas with high and medium rainfall in the northern and central districts, mainly the Hulah, Jezreel and Zebulon Valleys, the plains in Judea, Samaria and Galilee, and the lands bordering upon the mountains in these areas. This is the most common type of soil in Israel.

(2) *Calcareous soils*—in arid and semi-arid zones, mainly the

Jordan Valley and most of the Beth Shean Valley. This type of soil is also common in the hill regions. In the main this soil is deep in the valleys and superficial in the highlands.

(3) *Brown-red sandy soils*—principally in the Coastal Plain and in two large blocks in the north-western Negev and the Arraba respectively.

(4) *Clay-loam soils*—are to be found mainly in the area intervening between the sandy soils of the Southern District and the loess of the Negev. In this area there are also extensive areas of soil combining all three types—clay, sand and loess.

(5) *Loess soil*—in the Negev.

(6) *Peat soils*—in Hulah marsh area.

Where the various types of soil meet, intermediary or mixed types are common. These mixed soils sometimes differ from the base-rocks in colouring, composition and other qualities.

Saline soils do not constitute a specific category, as any type of soil may become saline as a result of an excess concentration of soluble salts. Blocks and even extensive stretches of saline soils are to be found in the Jordan Valley, the Arraba,[1] the Beth Shean, Zebulon and Jezreel Valleys and the Negev. The degree of salinity varies from some tenths of 1% to 10% and even 20%.

We shall discuss in brief some of the qualities of the main types of soils.

(1) Alluvial clay soils have been created by the weathering of hard chalk or basalt. The eroded material was transported by floodwaters into the valleys and plains. Some of this material remained on the rocks, where it forms the highland soil. These soils are commonly known as *terra rossa*, their colour being red or brown, but at times also black or grey.

The clay content is high, varying between 40% and 80%, rising in direct proportion to the depth of the soil, and retarding the percolation of water. The main chemical compounds of these soils are—silica (SiO_2) 30%–65%, oxides of iron and aluminium (Fe_2O_3–Al_2O_3) 15%–30% and chalk ($CaCO_3$) 10%–20%. Organic matter comprises approximately 1% and nitrogen 0·05%–0·12%. A higher percentage of organic matter (2%–5%) and nitrogen (0·2–0·4%) is to be found in the mountains.

The main agricultural crops on these soils are: (unirrigated) wheat, barley, durrha, maize, fodder, vegetables, grapes, almonds, figs; (irrigated) clover, maize, vegetables, grapes, citrus, bananas, apples and plums.

[1] 'ARRABA' in the Bible is synonymous for 'desert' (Isaiah li. 3). It is not a geographical proper noun but is used to denote the lowland of the Negev, which is devoid of trees or forest and is more or less similar to a 'steppe'.

(2) Calcareous soils are formed from eroded materials of soft chalk, washed down into the valleys.

The relative proportion of clay is lower than in the previous type (35%–45%). The chemical character also varies because of a different composition containing chalk and less silica (15%–40%). For this reason these soils are more permeable, despite the fact that they are classified as heavy and medium.

The reaction of the clay and calcareous soils is weakly alkaline.

Permeable calcareous soils provide a good bed for irrigated crops, particularly lucerne, clover, maize, vegetables, bananas, grapes and citrus (grapefruit).

(3) Brown-red sandy soils were formed by the weathering of chalky sandstone, locally known as 'kurkar', which is a brittle compound of chalk and sand. In these soils the sand content varies between 70% and 95% and that of clay between 5% and 30%. These are the lightest of Israel's soils. At greater depth the proportion of clay rises and frequently one metre below the surface a compressed, impermeable stratum, locally known as 'nazaz', which adversely influences deep-rooted plants, is formed.

Sandy soils are devoid of chalk and are fundamentally silica to an extent of 85%–95%. They require a supplement of lime. Their reaction is very close to neutral, being very weakly acid or alkaline.

This type of soil is excellent for citrus. Vines, almonds, bananas, vegetables and irrigated fodder also do well upon it.

(4) Clay-loam soils were formed by the weathering of sandy chalk-stone and can be classified into a number of categories, most of them as loams and some as sandy loams, clay-loams or clays, the clay content varying between 40% and 50%. The silica content (SiO_2) varies between 40% and 80% and chalk between 5% and 30%.

In the main the reaction is alkaline.

These soils are fertile and suitable for unirrigated cultivation of barley, wheat, durrha, sesame, maize, watermelons, vegetables, grapes, figs, olives, and to a lesser extent, under irrigation, for vegetables, fodder crops, vines and citrus.

(5) Loess soils developed from the silt-like material carried by winds from the desert regions and deposited on chalk hills. Their clay content is 10%–20% and silt 10%–35%. Silica is prominent in the chemical composition of these soils—50%–60%—while chalk comprises 12%–15%. The reaction is mainly alkaline.

It is difficult as yet to reach any definite conclusion regarding the fertility of loess soils, as they are to be found in an arid region, where only barley with a low stand is cultivated. Experiments conducted in recent years appear to justify the hope that given suitable irrigation various crops can be raised upon them. However, a

7

lengthier experience is necessary to test the composition of agricultural crops in the Northern Negev area.

(6) Peat soils were formed through the accumulation and decay of vegetation, principally papyrus, in the water of the Hulah marshes. Hulah peat belongs to the variety of low moor peat. The proportion of organic matter varies between 53%–77% and contains 35%–43% lignine, 8%–10% hemicellulose and 2%–4% cellulose. The reaction is very weakly alkaline.[1]

It is thought that following the conclusion of drainage works in the area, enabling the removal of excess water, the soil will be found suitable for the cultivation of vegetables and fodder. Another proposal envisages exploitation of the peat deposits for the production of organic fertilizer.

We have already stated that from the point of view of agricultural geography the most desirable zoning of the country should be into at least eighteen districts.

In Galilee there are in effect four distinct areas: Upper Galilee, the Hulah Valley, Western Galilee and Lower Galilee. The eight valleys —the Jordan, Beth Shean, Jezreel (east and west), Zebulon, Hefer Valleys, the Carmel Coast and the Arraba (in the Negev)—differ widely. Other distinctive districts are the Ephraim Mountains, Samaria, Sharon, the Lowland, the Southern District, the Judean Mountains and the Negev.

Upper Galilee

The northern section of the Upper Galilean highlands lies outside the borders of Israel; the lower section, however, which is part of Israel's territory, includes some of the highest mountains in the country. These mountains are composed of hard dolomite. The mountain range, which begins in the neighbourhood of the city of Safad, is known as the Mountains of Naphtali. This range constitutes

---

[1] (1) S. Ravikovitch, 'The Composition of Palestinian Soils', *Journal of Engineers and Architects in Palestine*, Vol. V, No. 6, August 1944. (2) S. Ravikovitch, 'The Aeolian Soils of the Northern Negev', *Agricultural Research Station Records*, V. 2–3 Israel, 1952. (3) S. Ravikovitch, 'The Brown-red Sandy Soils of the Sharon and the Shphela', *Agricultural Research Station Bulletin, Rehovot*, No. 55, 1950. (4) S. Ravikovitch, 'N. Bidner-Bar Chava. Saline Soil in Zevulun Valley', *Agricultural Research Station Bulletin, Rehovot*, No. 49, 1948. (5) A. Reifenberg, *The Soils of Palestine*, London, 1947.

the main watershed of streams flowing both to the Mediterranean and to the Jordan and Lake Hulah. It is a typical highland region devoid of high plateaux or extensive plains, and without large basins or wide valleys. The largest and prettiest valley is the Vale of Kadesh Naphtali.

Many springs are to be found in the smaller basins, forming streams which flow down the slopes, saturating the soil and producing a luxuriant vegetation. At one time the area was wooded. Today in this era of irrigated agriculture in Israel the unfavourable topographical formation constitutes a serious obstacle to the cultivation of larger areas. The shallower soils are suitable for winter grains, but despite the abundant rainfall and because of the inadequate depth of the soil, the moisture retained is not sufficient to nourish summer crops, which consequently must be irrigated.

Even the thriving orchards in this area, mainly apples, are in need of irrigation. Because of the unfavourable topographical conditions, only restricted areas of irrigable land suitable for fodder crops, which provide the basis for dairy-farming, are available. As against this, however, the rich pastures of these highlands are eminently suitable for sheep-raising. The luxuriant and variegated flora also constitutes a rich source of nectar for bees.

Agriculture in this area will be revolutionized in the near future, following the completion of a local irrigation project, leading water from the bountiful Malcha springs, by pipeline running north to south parallel to the main highway. Before long the hill settlements expect to obtain water for irrigation, though the excessively high cost will present a major problem for agriculture. A solution may be found if it should eventually be resolved to fix a standard price for water for irrigation purposes throughout the country.

Another difficulty with which the district has to contend is its exclusion from the countrywide electricity grid, as a result of which each individual settlement is compelled to maintain a separate plant for the generation of power.

The Hulah Valley

The Hulah Valley is a remarkably level plain and for this reason, as well as because of the natural obstacles blocking the mouth of the lake, extensive marshes have been formed. Irrigation plans envisage the removal of these obstacles and the reclamation of a large part of the lake. But even today, before the completion of the project, the valley benefits from the abundant and cheap water available. Topographical conditions in most parts of the valley, with the exception of the settlements on the slopes along its perimeter, are

9

extremely favourable. Here the soil is mainly of the alluvial clay. The land is fertile and all conditions for the development of irrigated farming exist. Dairy-farming, based upon various types of cultivated fodder and particularly lucerne, flourishes, while vegetable growing and irrigated orchards—mainly apples—are other thriving branches of local agriculture.

Sheep-raising is based upon the natural pastures of the foot-hills, while good conditions for the development of artificial pastures indicate favourable prospects for breeding beef-cattle. Industrial crops, too, including sugar-beet and ground-nuts, produce good results. Abundant supplies of water have stimulated the development of carp-breeding ponds.

Extensive, and often excessive, irrigation makes the improvement of the soil by drainage necessary.

Lower Galilee

The range of mountains links up with a series of hills towards the east above the Jordan Valley and Lake Kinneret. In a north-easterly direction lie the Mountains of Nazareth, broken up into many hills with a gentle slope. A chain of hills, a spur of these mountains, towards the north-west, separates the Vale of Zebulon from the Jezreel Valley.

To the north there is a second range, the Mountains of Turan, broken at the Vale of Beth Netupha, one of the larger vales of this country, for which a major role has been reserved in Israel's master irrigation plan. Two mountains, Givat Hamoreh and Mount Tabor, both south of the Nazareth range, are renowned in history.

The main watershed runs between these two mountains close to the city of Nazareth, through the Mountains of Nazareth from east to west. The larger streams of Lower Galilee are Wadi Malikh,[1] flowing into the Kishon, and the Chilazon which flows into the Naaman in the Valley of Zebulon.

Lower Galilee is not endowed with abundant water. Because of the mountainous terrain, characteristic of most of the area, and unfavourable topographical conditions, agriculture is necessarily extensive, based mainly on field crops producing only moderate yields. The olive, predominating over extensive areas, recalls the nature of Arab agriculture in the recent past. Jewish farmers do not favour the oil olive as the yield is low in alternate years while market-

---

[1] 'WADI'—an Arab word meaning a rocky gully resulting from soil erosion, dry in summer and flooded in rainy season. Larger wadis, which were drained or otherwise meliorated, are similar to vales.

prices do not cover the high costs of picking the fruit. No technical method of performing this task has so far been found. Stone-fruits and grapes will gradually oust the olive.

Favourable conditions exist for sheep-breeding, while the raising of beef-cattle may well expand as natural grazing is fairly abundant.

This district is included in the national irrigation plan, the execution of which will revolutionize local agriculture.

Western Galilee

There appears to be a certain correlation between altitude and rainfall in this region. In Western and Lower Galilee where altitudes are lower than in Upper Galilee, there is less rainfall, the annual average ranging between 600 and 700 millimetres (24–28 inches).

An interesting feature of Western Galilee is the two distinct types of soil which are to be found, a heavy, almost impermeable type bordering on light sandy soil. The Kurdani river, the waters of which are saline but can be used for irrigation on lighter soils, runs through the heavy soil area. It has been suggested that this water be mixed in conservation dams with sweet water obtained from borings and wells. Meanwhile the stream is being utilized for the development of carp-breeding ponds. Irrigated farming has not assumed extensive proportions in this area, though experience gained in recent years and conditions of soil, climate and water, have raised the question whether the cultivation of citrus, bananas and other sub-tropical varieties of fruit might not prove profitable.

The Jordan Valley

The Jordan Valley stretches from the point where the River Jordan issues from Lake Hulah in the north along the entire course of the river down to the Dead Sea in the south. This elongated valley, together with the Arraba, constitutes part of the great depression continuing via the Gulf of Akaba down to East Africa.

The Israel-Transjordan frontier runs through the northern section of the Jordan Valley (part of which is in Israel territory) and the Arraba, south of the Dead Sea. It is a great depression, comprising a number of distinct sections, such as the Genossar Lowland in the north, the Beth Shean Plain in the middle and the Jordan Plain towards the south. The Jordan Valley includes only the northern section of the Kinnarot Lowland, which entirely surrounds Lake Kinneret (the Sea of Galilee). The Kinnarot Lowland lies about 200 metres below sea-level and has a typically sub-tropical climate. For this reason only irrigated farming is engaged in. Despite the fact that

11

rainfall is low, water is abundant and cheap as it can be obtained in practically unlimited volume from the Jordan and the Yarmuk rivers.

Soils in this district are mainly of the deep calcareous type and are highly permeable.

This combination of high temperatures and cheap water, as well as the fact that calcareous soils provide an excellent bed for irrigated crops, has assisted in the development of a highly intensive agriculture. The chlorine content of the water of the Jordan is higher here than in the Hulah Valley, before the river enters the Kinneret, fluctuating between 120 and 300 milligrams to the litre. However, forty years of irrigated farming in the area have not resulted in any deterioration of the soil. Bananas and lucerne are the principal local crops. Fish-breeding ponds are also highly favoured, more so indeed than fishing in Lake Kinneret. Before these three branches of agriculture achieved their present pre-eminence, the pride of the valley was the tomato. Today, too, the Jordan Valley produces highly profitable early tomatoes and cucumbers. Lucerne, together with a large variety of other irrigated fodder crops, has stimulated the rapid development of dairy-farming. The lack of natural grazing on the other hand has retarded sheep-raising and cattle-breeding.

In addition to the banana and table grapes, the cultivation of subtropical fruits, especially dates, has been introduced.

In planning the future development of the area priority will have to be given to crops whose water requirements are more modest, for local farming is already menaced by a rise in the level of subterranean water and salinization of the soil.

An extensive drainage system and suitable rotation of crops to balance the use of water are essential to ensure the future of agriculture in this bountiful region.

The Beth Shean Valley

The Beth Shean Valley is a level plain below sea-level. Temperatures in the valley are generally high, though frost, causing much damage in banana plantations and tomato fields, is not unusual during the winter nights. During the summer evaporation is high, preventing the use of ordinary sprinklers, as the fine spray emitted does not saturate the highly permeable chalk soil, which requires large quantities of water. Most of the area consists of calcareous soils, the chalk content being higher here than elsewhere and exceeding 70%. A large part of these calcareous soils are saline—sometimes to a degree preventing cultivation. There is also brown soil in the district which is more suitable for cultivation.

12

Water for irrigation is supplied by the numerous springs at the foot of Mount Gilboa. The water of the larger springs, however, is saline and contains a high percentage of calcium carbonate ($CaCO_3$). But in addition to this, we encounter in this region the phenomenon of salinity of large areas which would normally cause serious agricultural problems.

These problems were solved by the establishment of carp-breeding ponds, this fish developing satisfactorily in saline water and enjoying especially the warm climate of this district.

The establishment of fish-breeding ponds in this region is fully justified even from the agricultural point of view, although limits should be set to an expansion of this branch of farming which requires large areas of land and a considerable amount of water.

In the light of these considerations this branch of farming will have to reckon with legitimate opposition from the Agricultural Department of the State, especially if the farmers of the district should emphasize the trend to extend fish-breeding over areas of good land and sweet water. Such a development indeed may react unfavourably upon the density of population in this border zone, which has favourable agricultural prospects.

Here, too, as in the neighbouring Jordan Valley, lucerne is the major irrigated crop. In spite of this the expansion of dairy-farming has been retarded because so far no breed of cattle has been satisfactorily acclimatized. The lucerne, accordingly, which flourishes in this district, is ground into meal, finding a ready market among cattle- and poultry-breeders in all parts of the country.

Other field crops, including winter grains, are all in need of supplementary irrigation to secure a yield equivalent to that obtained elsewhere on unirrigated soil. The major factor responsible is the inefficient and primitive methods of irrigation which cause salinization of the soil.

Vegetables can be raised successfully on suitable land with proper methods of irrigation, rational crop rotation and the eradication of pests and diseases. Potatoes grown here enjoy the important advantage of being the first in the market.

In regard to fruit-growing, this region is not among the most suitable; for example the date and the pomegranate grow better in other fruit-growing districts. Again both citrus and vines suffer from salinization of the soil, while the banana, which flourishes nearby in neighbouring Jordan Valley, has to contend here with two enemies —saline water and frost.

More recently a resolute attempt has been made to grow cotton on a considerable scale in the valley. The results have been highly encouraging, despite the fact that the sponsors of the crop shortened

13

the experimental period. The import of ginning machines will give cotton growing in this and other areas in Israel a powerful stimulus.

Eastern Jezreel Valley

This district borders upon Mount Gilboa. The slopes of the Mount on the Israeli side of the border are suitable for afforestation and for the development of frontier villages, whose farm lands will lie below in the valley. The soil is mainly of a heavy rich type of alluvial red-brown clay.

Surface water, supplied by the Harod spring, does not satisfy local requirements and irrigation needs are met mainly from borings and wells. The water of some of these wells is highly chlorine and is utilized to develop lucrative carp-breeding ponds. Most of the valley lies at an altitude averaging 60–70 metres above sea-level, with a rising gradient towards the west.

In 1922, when settlement work was first launched in the Eastern Jezreel Valley, most of its area was covered by fever-infested marshes. The Jewish settlers drained the soil, canalized the brook running through the valley and constructed a ramified irrigation system, administered by a District Co-operative Society. In the initial period the settlements were built on the banks of the Harod Brook, but after a number of years they were removed to a hillock to the west of the valley, facing Mount Gilboa. This removal to higher ground was dictated by both sanitary and strategic considerations. Serious agricultural problems ensued, however, arising mainly out of the need to distribute the various branches of farming, and particularly those dependent upon irrigation, between the valley and the hill, which lies at some distance away, and to which water must be pumped under pressure. As a result the valley has been developed in an interesting fashion. Just below the slope are the fields of green fodder. Closer to the settlements are the nurseries and the vegetable gardens, while higher up are the orchards. The main consideration in this planning of the farms has been the volume and cost of water available. The altitude of the particular branch of farming and its distance from the source of water is in inverse ratio to the volume of water it consumes.

One co-operative village situated in the valley has changed its farm plans twice because of these considerations. When the comparative advantage of dairy-farming dictated its proximity to the dwellings, the fields of green fodder were also moved, despite the increase in the costs of water involved. Following a decline in the price of milk the fodder fields went back to the valley where water is cheaper.

14

Today the plan is once again being altered with a view to concentrating these lands nearer to the farmyards.

Farming in this and the neighbouring district to the south has been based from the very beginning on field crops and dairying. In addition to irrigated green fodder, of which mention has already been made, winter grains—mainly wheat, barley and unirrigated fodder crops—are being successfully cultivated. Maize, a summer crop, has not done well here, apparently because of the lack of dew, though sorghum has been successful in recent years. Wintersome serves as an intermediary crop between clover in winter and sorghum in summer.

Excellent yields of beet, mainly for ensilage, are obtained.

All these crops, of course, provide the 'raw material' for dairy-farming, the central pillar of Israel's mixed farm economy.

In addition to poultry-breeding, another highly popular branch of agriculture in this district, sheep-raising has registered record achievements. It will be seen accordingly that the live stock branches of farming are firmly entrenched in the Eastern Jezreel Valley.

The grape was the first of the fruit crops to be developed in the valley, followed, after an interval of ten years, by citrus, and table-grapes and grapefruit which are now the principal fruits of the district.

Western Jezreel Valley

The western section of the Valley of Jezreel is a plain lying higher than its eastern counterpart. The altitude is between 25 and 50 metres above sea-level. For this reason rainfall here is higher than in Eastern Jezreel—600 millimetres annually as against the 400 millimetres of the latter. Adequate rainfall combines with a fertile soil, mainly of the alluvial clay type, to make the Western Jezreel Valley one of the most stable farming districts in the country, measured by the yield of winter crops. This is typical field-crop country and only in excessively wet years do the yields suffer. Drainage, however, can provide a solution to this problem.

To the Jordan Valley goes the honour of being the cradle of the collective settlement movement (the 'kibbutzim'). Western Jezreel Valley can claim similar merit in respect of the co-operative settlements (the 'moshavim'), the villages founded here over thirty years ago continuing to march in the van in all that pertains to agricultural, organizational and cultural development. Model farms are the rule in this district.

In other areas irrigation has constituted the basis for the development of agriculture. Here irrigation has enabled a reduction in the area of the farming unit and a higher density of agricultural population.

15

The ambition of the farmers of this part of Jezreel is to increase the area under irrigation and to devote themselves to an increasing extent to the development of industrial crops, such as ground-nuts and sugar-beet, which they are convinced will produce good yields.

The Kishon area, which is part of the Western Jezreel Valley, is less bountifully endowed for field-crop farming than the rest of the district. It is interesting to note that the few tracts under field crops in the Kishon area produce better yields in dry years than in wet. On the other hand summer crops flourish. The meagre success attending the cultivation of field crops and the good yields of summer crops are the result of excessive rainfall and heavy, poorly-drained soil.

An abundance of comparatively low-cost water obtained from not very deep wells facilitates the production of green fodder for the flourishing dairies, while the cultivation of vegetables is vigorously pursued. Vineyards and citrus groves are also cultivated.

Valley of Zebulon

The Valley of Zebulon is a level plain washed by Haifa Bay. Its shores are covered with sand, penetrating sometimes two kilometres inland. The marshes which at one time infested the valley were drained by Jewish settlers.

The valley is watered by two streams, the Kishon, which flows along its southern boundary, and the Naaman, crossing the valley from south to north, and influencing the character of the soil.

Annual rainfall is high and the floodwaters coming from afar fill the dry watercourses and create marshes on both banks of the Naaman along the coast. At some distance from the Naaman subterranean water can be struck at a depth of 15–30 centimetres. This water, however, is saline.

Three main types of soil are met with in this district—alluvial brown clay soils, grey sandy soils and saline soils of the type of solontshak (containing an excess of sodium salts, generally the chloride and the sulphate salts). The first two types are suitable for all varieties of irrigated crops and especially for fodder, which provides the basis for the extensive dairy-farms of the Valley of Zebulon. The saline soils are being improved by drainage.

Carmel Coast

The Carmel Coast extends from Mount Carmel near Haifa to west of Benyamina, its width never exceeding five kilometres along the entire 25 kilometres of its length. Thus it constitutes a narrow, level

16

strip of land, to which the proximity of the Carmel adds a dark grey chalky soil washed down from the mountain slope. The climate is similar to that of the coastal region, warm and humid in summer and mild in winter, thanks to the prevailing westerly winds from the Mediterranean. Rainfall averages 600 millimetres annually.

Until quite recently irrigation was not widely practised in the Carmel Coast district, and indeed it was only after the establishment of the State of Israel that it was included in an irrigation project, the first stages of which have already been completed. In 1953, 215 million cubic metres of water were supplied to the district.

Irrigation here serves to increase yields of seasonal fodder crops, lucerne (for meal) and ground-nuts, as well as for the cultivation of banana and citrus groves.

Special attention is being devoted to the development of dairy-farming in the co-operative villages of the coast.

In the few mountain farms in the district the cultivation of vineyards is an important branch of agriculture.

The Mountains of Ephraim

The Israeli highlands can be divided into the Mountains of Galilee in the north; the Mountains of Samaria (or of Ephraim) in the central district; and the Mountains of Judea in the south.

The Mountains of Ephraim contain more arable farm land than the neighbouring Carmel range, but the soil of the alluvial clay and calcareous types is shallow in the highlands and does not retain moisture for very long. Nevertheless, fairly stable yields of winter crops are obtained. Deciduous fruit and vineyards (table and wine grapes) do well here, and after suitable improvement of pastures the raising of sheep (mainly for mutton) and perhaps the fattening of cattle could prosper. All these, however, do not combine to create a sound viable basis for farming and it is necessary to supplement holdings with a plot of irrigable land in the neighbouring valley.

Samaria

From the vicinity of Zichron Yaacov in Samaria, southwards to Gaza, there extends the light and medium soil zone, of the red-brown sandy variety, good for citrus and particularly for oranges. The heavier soils of the alluvial clay type are more suitable for vineyards. For well over half a century the area has been developed as a centre of vine-growing served by the wine-cellars of Zichron Yaacov.

In recent years the district has produced large quantities of vegetables for the market.

17

The possibilities of cultivating fodder crops are limited, which accounts for the restricted dimensions of dairy-farming.

There is no surface water in Samaria, but this is compensated for by the abundance of borings and wells. A number of settlers, contrary to the provisions of national planning, have developed excessively larger fish-ponds. Water is too expensive here for this purpose while the land with a little irrigation is suitable for valuable crops.

Hefer Valley

The Sharon includes the entire area south of Samaria down to the northern approaches of Tel-Aviv. Sharon may be divided into three sub-districts—the Plain of Caesarea in the north, the Central Sharon or Emek Hefer and the South Sharon from Beth Lidd to Tel-Aviv. The Hefer Valley has excellent alluvial soil and bountiful sources of subterranean water at no great depth. Dozens of settlements of all types have been established here, the farms prospering on holdings no larger than 20–22 dunams in extent.

Green fodder irrigated with water that is not expensive has enabled the development of thriving dairy-farms. Recently it has been proved that sorghum can be successfully cultivated in this district.

Citrus groves constitute a branch of mixed farming here, while bananas have been introduced, though they are sometimes damaged by frost.

A characteristic feature of this area is the heterogeneity of the soils, resulting in excessive parcellation of the farm-unit—which in any case is not large. The evil effects of this fragmentation is felt particularly in the one-family farms.

The predominant type of soil in the valley is of the brown-red sandy variety, bordering upon sea sand in the west, and heavy clay soils in the north-east. This juxtaposition of different types of soil is typical of all Sharon; it must be recalled, of course, that Hefer Valley is part of North Sharon.

Sharon

The Sharon comprises a range of low hillocks beginning close to the sea and concluding near the lengthy plain of North Sharon.

The marshes that at one time infested the area and have since been drained were created by a combination of natural conditions— streams whose flow into the sea was blocked by the sand-dunes, the high level of subterranean water and inadequate drainage. Throughout the centuries reclamation of the soil has been a cardinal concern of the inhabitants. There is reason to believe, indeed, that the Falik stream, which runs through the district, is a large drainage canal excavated during the Roman occupation.

A number of streams water the Sharon, the biggest of them being the Yarkon, whose headwaters are the Rosh Ha'ayin springs. The Yarkon is also fed by a number of smaller streams and brooks before it flows into the Mediterranean. The Rosh Ha'ayin springs constitute the main source of Jerusalem's water supply. A typical feature of the Sharon landscape is its red sandy soil, which is excellent for the cultivation of citrus. Mixed and monocultural farms—the latter based upon citrus, poultry-breeding or vegetable growing—abound in this district. The settlements situated close to Tel-Aviv, the main consumption centre in the country, thrive on vegetable growing. Poultry-farming, too, is more highly developed here than elsewhere, one village in the district engaging solely in this branch of agriculture. Recently the breeding of water-poultry, mainly geese, has begun. The proximity of Tel-Aviv, which has stimulated the development of special types of farms in Sharon, is leading to the urbanization of the larger villages. This aspect will be dealt with elsewhere.

The Lowland

The Judean Plain, flanked by the Judean Highlands, and extending down to the Negev, is the natural continuation of the Sharon. It is an extensive area and comprises a number of distinctive geographical sections, including Yarkon area, the Lydda Plain, the Javneh Plain, the Northern, Central and Southern Lowland (or Philistia). Each of these sub-districts is distinguished by special natural conditions. Along the coast of the Mediterranean lies a belt of sand-dunes. East of the dunes extend ranges of hills, with intervening plains comprising a variety of soils, from heavy to light. The soil here is partly alluvial clay, partly red-brown sand, with clay-loam soils predominating, intermingled with various intermediary types. Many winter-streams, all of which dry up in the summer, cross the length and breadth of this district. Gravel and silt brought down from the mountain slopes fill the beds of the watercourses.

These streams, in the western section of the Lowland, form small plains, fertile because of the thick stratum of silt which covers them. In the vicinity of Ramleh–Lydda these smaller plains converge to form a single larger plain.

Despite the fact that there are few springs here, the entire lowland is extremely fertile. Annual rainfall fluctuates between 500 millimetres in the Tel-Aviv area to 350 in the Southern District. This is the second largest area in the country in point of area and has the highest density of population. It includes settlements of a great variety of types—cities, 'moshavot', suburbs, engaging in dairying, 'kibbutzim', 'moshavim', experimental stations, farm schools and transitional work camps ('maabarot').

Characteristic of this area are the extensive citrus groves and the vineyards whose produce is processed in the huge wine-cellars of Rishon-Lezion.

During the past year over 60 million cubic metres of water for irrigation were supplied to the Lowland, of which 10 million cubic metres were intended to encourage the cultivation of citrus and sub-tropical fruits. Dairy-farming has been intensively developed and calves bred by farms in the area are now being transferred to less developed districts, mainly to the younger settlements. At the same time it is already clear that the lands under lucerne and pastures must be extended to supply the needs of the livestock branches of farming.

The Southern District

This district is bordered by a line drawn from beyond Gedera to Bizaron, Beertuvia and Yad Mordechai in the west to Gath in the east, including all the settlements down to a line drawn from Dorot to Ruchama. The western strip close to the sea-shore has plenty of subterranean water. Towards the southern extremity of the district, in the vicinity of Nir-Am, lie the main boring fields, from which water was supplied to the first settlements founded in the arid Negev. The most common type of soil here is red sand giving way, towards the south, to clay-loam.

This is a region of groves, fodder and vegetables. Dry-farmed field crops do not produce satisfactory yields here, because of the low rainfall which decreases steadily as one travels south, and because of the excessive porousness of the soil. Lack of moisture in the soil also precludes the cultivation of summer crops. On heavier soils extensive vineyards are cultivated. Dairy- and poultry-farming are based upon irrigated fodder and barley crops.

Judean Mountains

The Judean Mountains can be sub-divided into three main sub-districts, namely, the Hebron Highlands, the Jerusalem Mountains and the Bethel Highlands. We refer here to that part of the Jerusalem Mountains which is situated in Israel territory, the altitude of which averages 750 metres, attaining a maximum of 885 metres. The countryside is bisected by scores of valleys, which throw the low, denuded dun hills, eroded by rain and wind, into stronger relief. At the foot of the slopes there are fertile valleys of accumulated silt. In ancient Israel this vicinity was occupied by a dense Hebrew population and the remains of an intricate system of artificial terracing, which fell into decay or was destroyed, are still clearly visible on the mountain slopes.

This section of the Mountains of Judea forms the eastern part of the Jerusalem Corridor. It is bounded by the Judean Lowland and the Transjordan frontier. Average temperatures are lower than in the plains and frost is usual in winter. Normally the air is dry. Rainfall is higher than in the Coastal Plain and averages 600 millimetres annually.

Natural conditions are favourable for the cultivation of fruit-trees. There is, however, little land suitable for the cultivation of field crops. Hillsides are usually terraced before the planting of trees. The soil on many of the slopes is too shallow even for orchards and large-scale afforestation is being carried out.

Stone-fruits, late varieties of grapes, figs and apples, do best in this district. In recent years special areas have been set aside for the cultivation of wine grapes by growers connected with the Rishon-Lezion cellars.

The settlements engage in dairy-farming, the fodder for which is purchased or in some cases cultivated by them on lands in the Lowland. Under these conditions it has been necessary to raise the yield of the cows, which indeed is the highest in the country and lends the dairy-farms in the district the character of suburban dairies. Poultry-farming also flourishes, though it is based entirely upon purchased feeding stuffs. But dairying, poultry-farming and fruit-growing do not constitute a broad enough basis for the villages in this area and the settlers are compelled to seek work outside their own farms. Two of the larger 'kibbutzim' have sought a partial solution to this problem by establishing rest-homes, but the economic consolidation of the mountain settlements remains a major difficulty.

More recently valuable experience has been gained in the cultivation of flowers and bulbs for export. This may develop and include cultivation of various herbs for pharmaceutical purposes. Another possible solution to the economic problems of these settlements is the establishment of workshops and industries, as in mountain villages in other countries.

The Negev

Negev means quite simply 'dry'. Annual rainfall in this district diminishes as one travels in a southerly direction and practically every kilometre southwards the decline is perceptible. A line drawn from Gaza through Ruchama and eastwards divides the area into the smaller semi-arid Northern Negev and the arid Southern Negev. In the northern section rainfall averages 350–400 millimetres (14–16 inches) annually, in the south only 25–50 millimetres (1–2 inches). Annual mean temperature is 20° C. in the north and 24° C. in the south.

At Migdal Ascalon the annual rainfall of 423 millimetres (17 inches) produces satisfactory yields; in Beersheba, only 25 kilometres south, the average does not exceed 227 millimetres (9 inches), while 25 kilometres further south annual rainfall is only 118 millimetres (5 inches).

Seasonal temperatures fluctuate sharply and the Negev is hot in summer and cold in winter. Changes in daily temperatures are frequent, while humidity is low. In the northern section of the Negev dew compensates to some extent for the inadequate rainfall, but further south this factor is non-existent.

Topographically it is possible to divide the Negev into four sub-districts:

(1) A flat plain extending mainly in a south-westerly direction at an altitude of about 100 metres above sea-level.

(2) A plain broken by low plateaux. This is the loess district constituting most of the area of the Negev. The altitude rises from west to east, over a distance of 90 kilometres, from 50 to 650 metres above sea-level (at Ras-al-Zuweira), after which there is a sharp drop to 392 metres below sea-level over a distance not exceeding 12 kilometres.

(3) The chalk hills in the south and south-east. This area has been excluded from all settlement plans because of the excessive softness of the chalk.

(4) The sands which cover an extensive part of the interior of the country and the coastal sands. The round and undulating character of these sands indicates constant movement.

Not so long ago the Negev was the exclusive habitat of nomad Bedouin who extracted a scanty livelihood from their flocks of sheep and cattle and small patches of barley sporadically sown in the desert, and averaging about 12–18 kilograms per dunam. Small beds of vegetables were also cultivated on the banks of the watercourses, where dykes were constructed to retain the moisture.

Another negative feature of the Negev, besides the lack of water, is the large number of dry watercourses and gullies, which make land uncultivable and create patches of saline soil, that become more frequent as one penetrates deeper into the interior.

The most hopeful prospect here is the development of various natural resources, but even for this purpose a minimal farming population is necessary.

Attempts have been made to bring water for irrigation purposes from the neighbouring districts but the volume available is very meagre and does not provide any solution to the Negev's major problem. Other plans, envisaging the piping of water from the Yarkon (execution of this project has already begun) and even from further afield, will be completed at a much later date and will give a stimulus to the repopulation of this desert area. (A special chapter is devoted to this aspect.)

We will conclude this chapter with a table setting forth the various agricultural branches in order of their financial importance within the sum of farm production during 1952/53, and giving also the main areas of cultivation in the order of their importance.

TABLE 2

Agricultural branches and main areas of cultivation

Branch of farming	% value of total farm production	Main areas of cultivation
(1) Field crops	20·6	Wheat: Jezreel Valley, Northern Negev, Southern District, Lowland, Vale of Zebulon, Beth Shean Valley and elsewhere
		Barley: Northern Negev (which comprises more than half of the entire area under this crop), the Southern District, Beth Shean Valley, Jezreel Valley, Sharon, Samaria and elsewhere
		Fodder: Southern District, Jezreel Valley, Northern Negev, Lowland, Sharon and elsewhere
(2) Eggs and poultry	18·0	Sharon, Southern District, Lower Galilee, Jezreel Valley and all other districts in Israel
(3) Milk and beef	17·4	Southern District, Western Jezreel Valley, Sharon, Samaria, Eastern Jezreel Valley, Hefer Valley, Upper Galilee and elsewhere
(4) Vegetables	16·0	Sharon, Southern District, Lower Galilee, Vale of Zebulon, Samaria, Northern Negev, Jezreel Valley and elsewhere
(5) Citrus	11·7	96·4% of the entire area under citrus in the State are concentrated in Judea and Sharon

23

Branch of farming	% value of total farm production	Main areas of cultivation
(6) Other fruit	6·6	Main varieties: Table Grapes: Western Jezreel Valley (18·3%), Jordan Valley (10·1%), Southern District (9·4%), followed by Galilee and the Coastal Plain Wine Grapes: Northern Negev (33·3%), Southern District (18·2%), Jerusalem Mountains (12·2%), Mountains of Ephraim (11%), Lower Galilee (9·5%), Western Jezreel Valley (7·8%) and elsewhere Apples: Upper Galilee (37%), Western Jezreel Valley (21%). Little in other districts Plums: Western Jezreel Valley, Jerusalem Mountains, Western Galilee, Mountains of Ephraim. 70% of the plum crop comes from these four districts Bananas: Jordan Valley (51%), Coastal Plain (37%), Southern District (6%)
(7) Fish	4·3	28% of the fish catch comes from marine fishing, 10% from inland lakes (Kinneret, Hulah) and rivers (Jordan) and 62% from fish-breeding ponds. The main areas of the ponds are Upper Galilee (34%), and Beth Shean Valley (26%), Zebulon Valley (14%), Coastal Plain (12%), Jordan Valley (10%) and Jezreel Valley (4%)

To complete the outline of the structure of Israel's farming it is necessary to mention that 4·2% of the production comes from various branches not mentioned above, while 1·2% is the value of the natural increase of livestock.

Meat marketed has been allocated between the poultry and the dairy branches in the ratio of 3 : 1.

The prevailing type of agriculture in this country is mixed farming, based upon the breeding of livestock and the cultivation of field crops (intended mainly to supply the livestock branches), and to a minor extent to provide cereals for human consumption. The cultivation of industrial crops is also engaged in to a minor extent. It is estimated that 50% of the total value of agricultural production is connected in one way or another with the livestock branches. It will be seen, however, from the above table, that the livestock branches

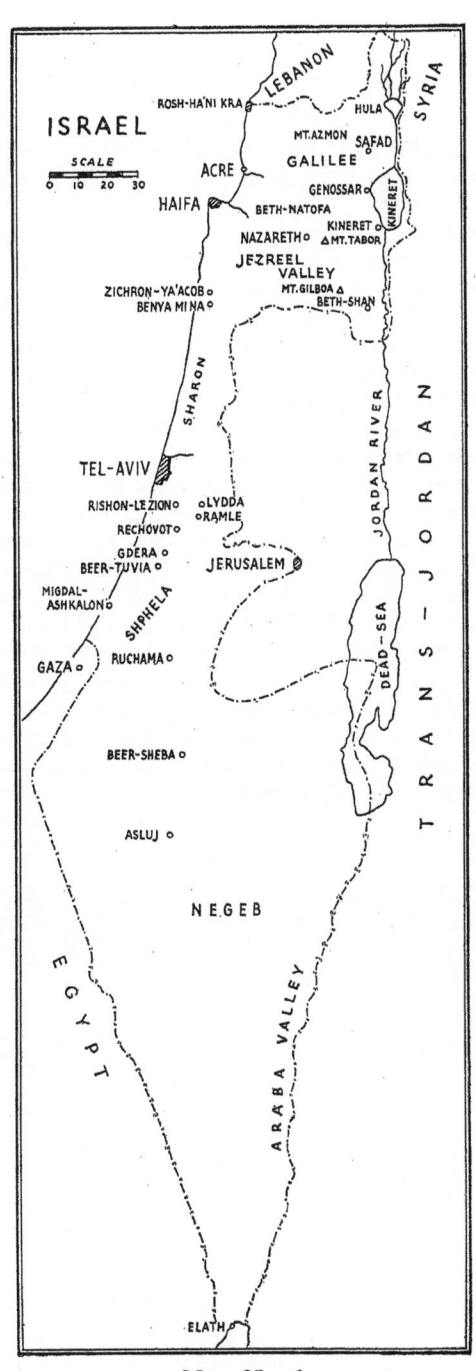

Map of Israel

are not necessarily concentrated in the same areas as those in which fodder and feeding stuffs are cultivated. Barley is sown in the Negev, for example, to supply the needs of the livestock branches in other parts of the country.

With the exception of citrus and carp-breeding ponds, which are restricted to a small number of districts, the main branches of farming are spread all over the country, though in varying proportions.

The picture given of Israel's agriculture in the above table might well change, especially in regard to the relative importance of various crops in each region, but the geographical distribution may be expected to alter little if at all.

Agricultural planning for the future puts the emphasis on citrus-growing, vineyards and other varieties of fruit. The importance of industrial crops is also expected to increase in view of the aid given to encourage the cultivation of sugar-beet, sugar-cane, sweet-sorghum (all for sugar production); ground-nuts and soya beans (for oil production); cotton and flax to provide raw materials for the local textile industry. Thus the field-crops branch will be required to supply raw materials for a number of industries, and for the production of animal proteins through the livestock branches of farming.

Vegetable-growing is the only branch of farming which appears to have reached its limits under present conditions. Other branches are distinguished by dynamic expansion, the pace being set by the progress of irrigation development works and the enlistment of new recruits to the ranks of the settlers.

2

LAND

A LMOST 80% of Mandatory Palestine (west of the Jordan), which covered an area of 27 million dunams[1] (27,024 square kilometres or 10,434 square miles), is included in the State of Israel. The remainder is occupied by the Arab Legion and the Egyptian forces.

Former Mandatory Palestine is now divided as follows:

State of Israel	20,873,469 dunams
Occupied by Arab Legion	5,891,234 ,,
Occupied by Egypt	299,544 ,,
Total	27,064,247 dunams[2]

461,000 dunams of the country's inland waters are within Israel territory and the remaining 209,000 dunams in the area occupied by the Arab Legion. Israel's inland waters include the Dead Sea—281,000 dunams; Lake Kinneret—166,000 dunams and Lake Hulah —14,000 dunams. The land area of Israel is thus 20,412,469 dunams. Data regarding the area suitable for cultivation is still meagre. A photogrammetrical survey of the country has been continuing for a number of years, and covers an area of 9,514,380 dunams from the northern border to co-ordinate latitude—0·60, 15 kilometres south of Beersheba. The Negev south of co-ordinate latitude 0·60 and east of co-ordinate longitude—140—of Israel's grid, is still unmapped.[3]

The area mapped has been divided into 21 districts, differing from each other in respect of soil and climatic conditions. These have been

---

[1] 4 dunams = about 1 acre.

[2] *Division of Land Survey.*

[3] N. Gill and Z. Rosensaft, *Soils of Israel and their land use capabilities. Summary of Soil Survey*, Part One, Ministry of Agriculture, Soil Conservation Service, Hakirya, 1955.

classified in four main categories, viz. the coastal area and adjacent valleys—3,148,090 dunams; the Jordan and Beth Shean Valleys—239,020 dunams; mountains and highlands—4,014,860 dunams; and the Northern Negev—2,112,410 dunams. The following data have been registered: texture and depth of soil, stoniness, water and wind erosion, floodwaters, marshes, salinity. Areas which are not suitable or available for agricultural purposes, such as built-up area, mounds, ruins, etc., have been demarcated.

In close collaboration with the agricultural planners the potentialities of the soil have also been determined: unirrigated land, land for irrigation, pastures, land for commercial afforestation and 'destroyed' land. The last category is also suitable for afforestation, though not on a commercial basis.

It is interesting to note that the area suitable for unirrigated farming has been placed at 4,613,580 dunams, and that for irrigated farming at 5,203,690 dunams. These figures may at first appear paradoxical. This is explained by the fact that there are lands which can only be cultivated under conditions of artificial irrigation. Without water no sort of tillage is likely to succeed. These areas are mainly concentrated in the Northern Negev.

An area of 3,016,290 dunams has been defined as suitable for pasture. Should, however, all suitable areas be brought under irrigation, part of the pasture land, which cannot otherwise be classified as agricultural land, would also be watered. In this case natural pastures would cover no more than 1,939,020 dunams.

According to these estimates the maximum area regarded as suitable for cultivation is identical with that which can be irrigated, namely, 5,203,690 dunams. This constitutes 54·7% of the entire area surveyed, or 24·9% of the territory of the State.

The following table compares the unit of land per person in Israel with that in a number of other countries.

TABLE 3

Units of land in different countries (dunams)

Country	Total area per capita	Agricultural land per capita	Average farm-unit
Australia	95·0	23·0	1147
England	4·8	3·7	359
United States	52·0	9·0	852
Belgium	3·5	2·2	19
Denmark	10·0	7·4	153
Holland	3·0	2·3	57
Norway	100·0	3·2	30
Israel	10·4	2·6	37

Of course, at best, only a very general idea of the average unit of land per capita in other countries and in Israel can be obtained from this table. It must constantly be borne in mind, that these statistics are governed and qualified by agricultural methods, rainfall, quality of equipment, the rotation of crops, the main crops, use of fertilizers, the average yield per unit of land and the like.

The figures in the above table are derived from a world agricultural census, conducted in 1950, by the Food and Agriculture Organization.[1] Those of Israel, however, are based upon agricultural development plans, as the figures for 1950 were not representative. Since then the area under cultivation has been stabilized, but the crops cultivated, as well as the potentialities of the soil, are changing, as a consequence of the rapid increase in the area under irrigation.

In respect of the size of the unit of land, Israel is closest to Belgium and Holland. The growth of population in these countries, as in Israel, underlines the problem of increasing the agricultural yield. In Israel, solution of the question of food production for a growing population will depend primarily upon the expansion of the area under irrigation. It will also prove necessary to bring new land under the plough. As already stated, the survey does not include land south of co-ordinate 0·60, 15 kilometres south of Beersheba. In the immediate future there is no need to conduct the necessarily expensive survey of this area. Should, however, the two million dunams of the Northern Negev be irrigated, it is very likely that this border will be pushed further southwards, provided, of course, that the water for irrigation is available.

CLASSIFICATION OF THE SOIL

The normal method of classifying land is by determining categories of agricultural and non-agricultural land. The former category includes areas suitable for woodlands, as well as pastures.

According to the survey referred to there are 8,739,610 dunams of agricultural land in Israel, classified as follows:

Cultivable land	4,613,580 dunams
Pastures	3,016,290 ,,
For commercial afforestation	878,100 ,,
'Destroyed' land and land for non-commercial afforestation	231,640 ,,

We have already noted that given irrigation the cultivable area can

[1] *Food and Agricultural Statistics Monthly Bulletins,* for the years 1951 and 1952, where Census Results of 1950 were published.

be increased to 5,203,690 dunams, mainly at the expense of pasture land. In such an eventuality the area of pasture land will be reduced to 1,939,020 dunams.

Pasture land includes highlands and hillsides with less than 50% of stones, the gradient of which is under 35 degrees.

The area for woodlands is situated in the districts with an annual rainfall exceeding 350 millimetres. The woodlands can be extended at the expense of the pasture land.

In regard to the agricultural land the question of their classification into marginal, sub-marginal and super-marginal categories—in other words lands whose yield is equal to, less than or more than the costs of production—remains to be settled.

It can be stated with confidence that all irrigable lands are of the super-marginal category. Citrus, three-quarters of the yield of which is intended for export or for the manufacture of conserves or concentrates, is cultivated upon this category of soil. Given suitable methods of cultivation even a medium yield gives a good return, as long as there is a market for the fruit overseas.

In general a satisfactory return, more than covering the cost of production, can also be obtained in local markets for irrigated crops, including varieties of fruit, vegetables, potatoes and fodder, etc. Barley, which serves as the main feeding stuff for livestock and poultry, is an unirrigated super-marginal crop. Only wheat is a marginal crop in certain areas—and even sub-marginal, where rainfall is insufficient and erratic.

It has, however, already been stated that classification of the soil must proceed hand in hand with cultivation. There may be changes in the categories defined from time to time, as a result of changes in methods of tillage and new production needs.

The survey has revealed 949,000 dunams of land which are not suitable for agricultural purposes. These have been classified as follows (on opposite page).[1]

The restricted area of land in the country and the high standard of living which farmers justly wish to achieve and maintain are the major motives underlying the intensification of farming methods and the widespread use of irrigation.

This trend is reflected in the constant process of reducing the area of the farming unit.

We shall examine more closely changes in the extent of the farm-unit in a number of settlements in various districts over the past fifteen years.

[1] See Note 3, p. 27.

Hillsides	232,000	dunams
Built-up areas	448,000	,,
Lakes	183,000	,,
Rivers and watercourses	44,000	,,
Ruins and antiquities	27,000	,,
Borderlands (unsurveyed)	15,000	,,
		949,000 dunams
For afforestation	878,000	,,
Pasture land (under all conditions)	1,940,000	,,
Existing fish-ponds	34,000	,,
		2,852,000 ,,
Land cultivable (under all conditions)		3,239,000 ,,
Land cultivable (after amelioration)		
Shifting sand-dunes (irrigable)	307,000	dunams
Land in the North Negev (irrigable) or natural pastures	1,077,000	,,
Saline marshes	12,000	,,
High subterranean waters	30,000	,,
Peat	32,000	,,
Land requiring heavy de-stoning	114,000	,,
Land requiring very heavy de-stoning	702,000	,,
Extensive orchards (in the Highlands)	200,000	,,
		2,474,000 ,,
Total		9,514,000 dunams

34% of the lands surveyed are suitable for agricultural purposes under all conditions.

26% can be rendered cultivable by varying degree of amelioration and improvement.

30% are suitable for afforestation, pasture and existing fish-ponds.

10% are already occupied for non-agricultural purposes.

UNIT OF LAND

The restricted cultivable area has naturally influenced the size of the farm-unit, and the area of land placed at the disposal of each agricultural family.

Under the near-feudal conditions reigning among the Arabs in this country prior to the establishment of the State of Israel, there existed a small class of wealthy 'effendis'[1] who owned extensive tracts of land. These effendis, however, did not develop large-scale farming estates similar to those cultivated in other countries, but preferred

[1] A Turkish title given to members of the upper classes, mostly senior government officials.

to lease their lands to tenants. Arab smallholders occupied only a few dunams of land, medium peasants some dozens of dunams and, at very most, 100–150 dunams.

Nor have large estates developed in the Jewish sector. Settlers in the colonies, established in the last quarter of the nineteenth century, were granted larger holdings, amounting to 200, 250 and even 300 dunams, based upon the cultivation of field crops. But the number of such colonies was not large. Today the situation even in these settlements has undergone a radical change and the unit of land has been reduced. Even in citrus-growing, which attracted private investors, the groves exceeding 100 dunams in extent account for only 3% of the total.

The general tendency, accordingly, is in the direction of a decrease in the unit of land, concurrent with an increase in the farming population and in the area under irrigation.

A study of the changes in seven settlements, situated in various districts, given in Table 4, will facilitate analysis of this trend.[1]

The cultivation of land under field crops in outlying areas is a general feature of the older settlements. This is done sometimes upon their own initiative and sometimes at the request of the authorities, who are interested in bringing unoccupied land, awaiting permanent settlement, under the plough.

The substantial increase in the unit of land cultivated upon a temporary basis in 1952 was the result of such arrangements reached between a large number of settlements and the various agricultural and settlement authorities. Extensive areas in the Northern Negev at a distance of 200, 300 and even 400 kilometres from the original settlement, were cultivated by veteran settlements. When, however, these lands are transferred to new settlers, the older settlements naturally discontinue cultivation. Thus it may be assumed that in the next few years these lands will no longer be temporarily cultivated. On the other hand there are settlements whose land unit was originally small, without any possibility over a lengthy period of enlarging the holding. In recent years the holdings of these settlements have been supplemented by abandoned lands placed at their disposal. The two last settlements—in the Hefer Valley and the Southern District, respectively—listed in the above table enter this category. It will be seen that their land unit has grown despite the increase in population. The case in regard to the Jordan Valley settlement is similar. The problem of this settlement, however, which had also to wait for

[1] The basic data were obtained from the yearly balance sheets of the settlements and from the annual reports of the Audit Union for the Workers' Settlements.

TABLE 4

Changes in land units in 7 'Kibbutzim' (dunams)

	P	1937 T	I	P	1947 T	I	P	1952 T
Settlement in Upper Galilee								
Per capita	21·3	—	2·3	12·4	—	1·1	10·6	9·7	1·1
Per unit	56·6	—	6·1	46·9	—	4·2	42·7	39·8	4·5
Settlement in Jordan Valley								
Per capita	5·1	—	2·5	6·2	2·1	4·1	5·8	9·6	2·9
Per unit	16·5	—	8·0	22·6	7·8	15·3	22·6	37·3	11·3
Settlement in Gilboa District								
Per capita	14·5	8·0	2·8	14·9	—	1·7	10·7	20·5	2·1
Per unit	47·0	26·0	9·1	53·1	—	6·0	41·2	78·9	8·0
Settlement in Western Jezreel Valley								
Per capita	24·4	—	0·2	11·8	1·8	1·5	7·9	8·4	1·5
Per unit	83·0	—	0·6	47·3	7·3	6·1	31·3	33·3	5·8
Settlement in Zebulon Valley								
Per capita	15·2	6·4	1·6	9·1	5·6	4·1	7·1	17·4	2·7
Per unit	43·8	18·4	4·5	33·9	21·0	15·3	27·6	67·1	10·5
Settlement in Hefer Valley								
Per capita	3·4	2·2	1·4	3·1	0·6	1·5	3·1	4·2	1·1
Per unit	8·8	5·8	3·6	11·8	2·2	5·8	12·6	16·8	4·4
Settlement in Southern District								
Per capita	2·1	—	1·5	1·9	1·6	0·9	5·3	6·9	1·3
Per unit	5·7	—	4·1	7·2	6·2	3·6	22·1	28·7	5·4

Note: Unit means the area of land at the disposal of two adults or two heads of a family. The number of farm-units has been obtained by totalling the number of settlers, both men and women and intending settlers (in the collective settlements), and dividing by two.
P=Land permanently held; T=Land temporarily held; I=Irrigated land.

many years for a supplementary grant of land, was solved by the purchase of a tract in the Zemach area prior to the outbreak of the War of Liberation.

In all settlements the gross area under irrigation has been extended. However, in some settlements a relative decrease has taken place as a result of expansion in population and the number of farm-units. The table overleaf gives the changes in population, number of farm-units, permanent land-unit and irrigated unit, in all seven settlements dealt with in the preceding table, expressed in terms of an index (1937 = 100).

TABLE 5

Changes in population, farm-units and land-units
in seven 'Kibbutzim'

Settle-ment number	1947				1952			
	Popula-tion	Units	Land-unit P	I	Popula-tion	Units	Land-unit P	I
1	185	130	107	90	238	158	119	118
2	153	132	186	248	198	164	225	232
3	167	152	171	100	204	172	151	151
4	225	187	108	1838	334	282	108	2610
5	162	125	96	424	205	155	96	360
6	244	167	225	266	409	262	375	316
7	180	128	163	112	230	152	593	201

A common feature of all seven settlements is the growth in population, the relative increase ranging from 198 to 409. The relative increase in the number of units ranged between 152 and 282. The permanent land holding of four settlements did not change virtually over a period of ten years, and in three it changed, if at all, only during the past five years.

Two settlements received most of their land after the War of Liberation, as they were unable to solve their land problem previously.

Settlement No. 5, whose land holding remained unaltered for fifteen years, registered a more than threefold expansion of the irrigated area. Settlement No. 3 had no water at all for irrigation purposes, and engaged in dry farming, as rainfall in the Western Jezreel Valley is sufficient to ensure satisfactory yields. Fruit-growing in this district was also unirrigated. The 'Mekorot' Corporation brought water to Western Jezreel at a later stage, which explains the high relative increase in the area under irrigation.

These seven settlements, situated in different districts, can be regarded as representative and indicate three major trends: (a) an increase in population and number of farm-units; (b) a steady decline in the size of the unit of land, despite the fact that in the majority of cases the gross holding was enlarged; (c) a contraction in the size of the irrigated unit in relation to the land-unit and number of workers, notwithstanding a constant expansion of the irrigation area throughout the State.

A similar process can be noted in the 'moshavim'. The following table facilitates a study of this process in four 'moshavim' situated in different areas. The figures refer to 1952.

With the exception of the 'moshav' in the Southern District where the area under irrigation more than doubled, the rate of growth of both population and irrigated area was more or less similar.

LAND

TABLE 6

Changes in population and irrigated areas in four 'Moshavim'

	Population (Index 1939 = 100)	Under irrigation (Index 1939 = 100)
(1) 'Moshav' in Gilboa District	120	147
(2) ,, ,, Western Jezreel Valley	89	149
(3) ,, ,, Hefer Valley	126	174
(4) ,, ,, Southern District	104	222

In regard to the number and size of units there were no substantial changes, in view of the fact that from the very outset each family was allocated its permanent holding. Sometimes members of the second generation of the village or new settlers develop farms on the reserve lands but such changes do not alter the general picture to any considerable extent. In many cases sons remain on their parents' farm even after marriage, because as a result of the intensification of cultivation and the increase in the area under irrigation more labour is required.

It is also only natural that the population of the 'kibbutzim' should grow more rapidly than that of the 'moshavim', but the differences in the increase in the area under irrigation are insignificant. In the 'moshavim' too, the trend is in the direction of a greater density of population.

CITRICULTURE

In view of the importance of citrus-growing in Israel's agricultural economy, it is of interest to classify the groves according to their size. This classification for 1951/52 is given in the following table:[1]

TABLE 7

Size of citrus groves

Size of grove (dunams)	No. of groves	% of total
3–4	660	15
5–9	1,450	32
10–19	1,210	27
20–29	470	10
30–49	350	8
50–99	240	5
100 and above	120	3
Total	4,500	100

---

[1] Citrus Control Board, *Annual Report for 1951/52*, Tel-Aviv, 1952, p. 2.

It will be seen that 74% of the total number of groves did not exceed 19 dunams in extent. Even this branch of farming, which traditionally is the preserve of the 'planter' class, in Israel does not enter the category of 'latifundia'.

CHANGES IN LAND-OWNERSHIP

The basic principles of Government policy approved by the Israel 'Knesset' on 11th March 1949, contains the following significant clauses defining the objectives of agricultural development policy:

'The rapid and balanced settlement of under-populated areas of the State and prevention of congestion in urban centres.' (a)

'The implementation of irrigation projects in the valleys and the Negev plains, afforestation of highlands, drainage of swamps, improvement of soil and the development of agriculture in all parts of the country.' (c)

'Nationalization of water and natural resources, of waste lands and of services essential for the defence of the State. No expropriation or confiscation will be made without due compensation.' (d)

'Compulsory development of neglected lands, either by transfer to the Development Authority of the State, or by the imposition of special taxation.' (e)

'The rehabilitation and expansion of citrus-growing, improvement of methods of cultivation and assistance in marketing the fruit abroad.' (j)

The development of agriculture is also mentioned in other clauses which deal with investment of capital, labour, vocational education, increased production, etc.

This programme, however, does not speak of contemplated patterns of land ownership, excepting for the brief mention of 'the nationalization of land in waste areas' already cited.

The need has not yet arisen to formulate such a policy. In this respect, indeed, Israel is unique, for without any official declaration or act of nationalization, most of the land is actually in public ownership—of the State or of a public land institution. Of the 20·4 million dunams of land in this country, 1,159,000 dunams are privately owned, 625,000 dunams by Jews and 534,000 dunams by Arabs. Thus less than 6% of the total is in the possession of private individuals. Of the remainder, 3·3 million dunams are owned by the Jewish National Fund (only a small percentage of these lands have not yet been transferred in the Land Registry) and 141,000 dunams by the Palestine Jewish Colonization Association. The balance belongs to the State, through its agencies, the Development Authority and the Custodian of Absentee Property. The Palestine Jewish Colonization Association acquired a total of 520,000 dunams of land in Israel, but has transferred most of its holdings to settlers. The pattern of land-ownership in Israel is reflected in the following table:

Government and Government agencies	77·3%
Jewish National Fund	16·3%
Palestine Jewish Colonization Association	0·7%
Private Jewish ownership	3·1%
Private Arab ownership	2·6%
Total	100·0%

UNCULTIVATED LAND

Among the consequences of the War of Liberation were the abandonment by the former cultivators of large areas of land, neglect of orchards and groves and wastage of water resources. In keeping with its policy of stimulating agricultural production the Government adopted a series of measures to prevent destruction of orchards and irrigation installations. The Minister of Agriculture was given emergency powers to ensure proper cultivation of lands that had been abandoned and left uncultivated. The 'Emergency Regulations for the Cultivation of Uncultivated Lands and the Utilization of Unexploited Water' were promulgated in 1949, authorizing the Minister to take these lands into his possession in order to ensure proper cultivation. The Minister may cultivate these lands, or transfer them to others for cultivation. The owners of the lands are entitled to receive any net land rental paid by the cultivators to the Ministry of Agriculture, but cannot claim compensation or any other payment from the cultivator. The entire yield of uncultivated land belongs to the cultivator.

In keeping with these regulations the Minister has leased these lands upon a seasonal basis to various cultivators.

ABSENTEE PROPERTY

In 1950 the Absentee Property Act was passed by the Knesset. Under the terms of this Act an 'absentee' is defined as

'a person, who between the period beginning 29th November 1947 and the day when it shall be proclaimed that a state of emergency no longer exists was the lawful owner of property in Israel territory and at any time in this said period was (a) a citizen or subject of Lebanon, Syria, Egypt, Saudia, Transjordan, Iraq or Yemen or (b) was living in one of these countries or in that part of Palestine outside Israel territory or (c) was a citizen of Palestine and left his normal place of residence in Palestine for some place outside Palestine, before the 1st September 1948 or for a place in Palestine that was at the time occupied by forces that sought to prevent the establishment of the State of Israel and conducted warlike operations against the State after its establishment'.

This Act provided for the appointment of a council and a Custodian of Absentee Property.

The Custodian registers all absentee property, which he holds himself or which is held by others with his consent. He may make any payments or investments he thinks necessary for the maintenance of the property he holds. He may sell or lease the property, wholly or partly.

The Custodian in fact administers various interests of absentee owners, including the cultivation of groves found suitable for the purpose. In many cases the Custodian has leased the property in question to interested parties.

The Act, however, limits the powers of the Custodian in respect of the sale or long-term lease of any assets as follows:

'Should the asset be immovable property the Custodian may not (1) sell or otherwise dispose of the title to the asset. But if a Development Authority be established by an act of the Knesset, then the Custodian may sell the asset to that Authority, at a price no less than its official valuation. (2) lease the property for a period exceeding six years, other than to the above-mentioned development authority. . . .'

The Development Authority is thus authorized to buy, sell or transfer property. In regard to land the Act recognizes a special category of 'land transferred for public use', which is defined as being 'immovable property which is not urban land, or urban land which the Government has declared, for the purposes of this Act, intended for the construction of housing for immigrants, popular housing or for the purposes of development.

'The Development Authority may not sell land transferred for public use or otherwise transfer title to such land, other than to the State, the Jewish National Fund, to any local authority, or to any institution approved by the Government, which may be set up under this clause as an institution for the settlement of landless Arabs, the title to land so acquired, may not be transferred other than to one of the bodies enumerated in this clause, with the consent of the Development Authority [(3) (4) (a)].

'The Development Authority may not sell landed property, which

38

is not land transferable for public use, without first offering it to the Jewish National Fund, and after the Jewish National Fund has refused to acquire such land within a term fixed by the Development Authority [(3) (4) (*b*)].'

THE JEWISH NATIONAL FUND

The main function of the Jewish National Fund since its foundation at the beginning of the present century has been to acquire land in Palestine on behalf of the Jewish people. Prior to the establishment of the State of Israel the Fund had purchased and otherwise acquired 942,100 dunams in all parts of the country.

Vast educational and organizational effort as well as large sums of money were invested over two generations in carrying out this function. The Jewish settlements, whose establishment throughout the country was made possible by these extensive acquisitions of land, constituted the nucleus of the Jewish State. Following the creation of the State the Jewish National Fund has continued to exist as the agency for the transfer of landed property in Israel to the ownership of the Jewish people.

Through the agency of the Development Authority large blocks of land which can be exploited for immediate settlement are being transferred to national ownership through the Jewish National Fund. The first series of such transactions provided for the transfer to the Fund of 1,092,000 dunams for a sum of £I.18 million paid into the State treasury.

The transfer of these lands and the work of settlement upon them continued for a period of two years, after which it was found necessary to transfer other blocks of land to the Jewish National Fund.

In 1951 a new agreement was signed between the Jewish National Fund and the Development Authority providing for the transfer of 1,245,380 dunams of abandoned lands. The lands were utilized for the establishment of 159 new villages as well as for supplementing the land holdings of 182 older villages. A portion of these lands was held in reserve for other new settlements, while smaller tracts were placed at the disposal of agricultural schools, housing projects and afforestation work.

When the transfer of these lands in the Land Registry is concluded, the total holding of the Jewish National Fund will aggregate 3,329,000 dunams as compared with 942,000 acquired prior to May 1948. This represents an increase of just over 250%.

The following table (overleaf) reflects the relative increase in the land-holding of the Jewish National Fund in the various districts of the country (1947 = 100).

Galilee	300	Judean Coast	609
Valleys	173	Jerusalem Corridor	865
Haifa District	258	Judean Lowland	1,523
Sharon	298	Northern Negev	474

This table shows that the greatest relative increase was in the Lowland and the Jerusalem Corridor. In each of these districts the former holding of the Fund was 17,000 dunams as compared with 259,000 dunams and 147,000 dunams, respectively, in 1954. In the Northern Negev the Fund held title to 95,000 dunams against its present holding of 450,000 dunams.

The process of redemption of the land in this country by Jews is faithfully reflected in the activities of the Jewish National Fund over the past half century. These activities can conveniently be divided into four main periods;

(a) The initial period lasting twelve years, from the foundation of the Fund in 1901 to the outbreak of the First World War. In 1914 there were 420,600 dunams in Jewish ownership of which no more than 16,400 dunams were the property of the Jewish National Fund. Upon this area the first settlements sponsored by the Zionist Organization were established. The bulk of other Jewish owned land was held by the Palestine Jewish Colonization Association (P.I.C.A.).

(b) The second period begins with the issue of the Balfour Declaration in 1917 and the reopening of the Palestine Land Registry in 1920, and continues down to 1940. Land in the possession of the Jewish National Fund increased to 532,000 dunams (in 1941) out of a total of 1,604,800 owned by Jews.

(c) The third period begins with the promulgation of the Land Ordinance by the Mandatory Administration in 1940 and lasts until the proclamation of the State of Israel. Despite the rigorous restrictions officially imposed upon the acquisition of land by Jews the holding of the Fund increased by 400,000 dunams in the course of these eight years.

(d) The fourth period covers the years following the establishment of the State. The removal of all restrictions upon the acquisition of land by Jews opened up new vistas for the Fund which could now embark upon the realization of its original purpose—the transfer of the Land of Israel to the ownership of the People of Israel.

THE ARAB VILLAGE

The system of land tenure in the Arab villages of Israel, like that obtaining in the country generally, has not been dislocated by any legal change.

The Agricultural Census conducted in the year 1950 revealed that

78·5% of the land in the possession of Arabs was privately owned, the remaining 21·5% being occupied on leasehold. 95·6% of the farms were on privately owned land. 30·5% of the cultivated land was partly leased and partly privately owned. 4·4% of the total area of the Arab farms was leased. Only farms, the area of which exceeds five dunams, are included in this context.

An investigation conducted by a Mandatory Government Commission in 104 villages in 1930 revealed that 68·2% of the land was cultivated by owners and the remaining 31·8% by tenants.

There has been a slight increase under Israel administration (as compared with the Mandatory regime) in favour of owner-cultivated land. This is probably because the lands of the larger landlords who left the country during the War of Liberation, now held by the Custodian of Absentee Property, are no longer cultivated by Arab tenants.

Considerable changes in the Arab system of land tenure may be expected as a result of environmental influences and the progressive agrarian regime in Israel.

NATIONAL OWNERSHIP OF THE LAND

The urgency of the need to bring Jews to Israel and to settle them on the land produced a system of national ownership of the land. This system continues unchallenged to this day and no alternative system is considered feasible.

Unlike many other countries Israel does not suffer from any agrarian problem, necessitating reforms in the system of land tenure. The land, as already stated, is mainly held by the Jewish National Fund—a national body—and the Government of Israel. Thousands of families occupy national land and engage in farming, without giving a moment's thought to who holds title to the land.

3

WATER

I N the early months of 1946, when the Anglo-American Commission of Inquiry was conducting its investigations in this country, Dr. H. Weizman, then President of the World Zionist Organization, who was to testify before the Commission, requested the Agricultural Workers Organization to prepare an estimate of the water resources of the country. He specifically requested a *minimum* estimate. A small delegation met Dr. Weizman to try to persuade him to quote a *maximum* figure, or at least some estimate between the maximum and minimum. The late President smiled and declared, 'Only twenty-five years ago I toured the country in the company of the two top-ranking experts of the time. They both insisted that not a drop of water could be found in Eretz Israel. Today your quarrel is whether we have four billion cubic metres or two and a half billion cubic metres. Today I can permit myself to be a minimalist. When we have the last drop of that two and a half billion I shall join those who say we have four billions.'

From the very earliest period, indeed, Palestine has been notorious for its dry climate. Its surface water did not attest to any bountiful water resources. Neither the Jordan nor the Yarmuk, nor the Yarkon —modest streams all three—are comparable to the mighty Nile, the Tigris or the Euphrates. Palestine never possessed a water system like that of ancient Egypt or Mesopotamia. Cisterns for the conservation of rainwater, in the hills, and pools elsewhere were common in this country throughout its history. The pool and the tunnel excavated during the reign of Hezekiah provided water for drinking purposes and not for irrigation.[1] Later, under the Roman conquest, a number of aqueducts, the ruins of which still stand, were con-

---

[1] ". . . and how he made a pool and a conduit and brought water into the city . . ." (2 Kings xx–xxi).

structed. The inhabitants of the country constructed canals to exploit the springs.

When Jewish immigration into the country started towards the close of the nineteenth century, no proper water system was to be found in Palestine. The new settlers introduced artesian wells, pumps, motors, filters, pipes and faucets, after they had suffered for years from transporting water in casks, buckets and petrol cans to their homes, from the communal well. Water shortage prompted a more energetic approach to the problem. During the thirties extensive water prospecting was conducted, and teams of geologists, geophysicists, hydrologists and technicians were enlisted. Both the settlement authorities and the settlers themselves assessed the future of agricultural development in terms of water for irrigation.

This drive for water was further fostered by attempts made by both the Arabs and the Mandatory administration to hinder the work of Jewish settlement. Pressure was exerted by a series of Commissions of Inquiry and the reports of various experts who sought to prove that there was no room for the co-existence of both Jews and Arabs in Palestine. The Jews counter-argued that the discovery of water and wider use of irrigation would in effect increase the cultivable area, as every irrigated dunam could yield as much as four or five unirrigated dunams, thus making room for additional settlers.

By the early forties the view that water was available was already subscribed to by recognized experts, and various estimates of water resources were published.

The main source of water in Israel is, of course, rainfall.

Six billion cubic metres of rain fall in Israel every year. In addition the Jordan brings floodwaters from beyond Israel's borders. A high proportion of this rainfall, some 70%, is immediately lost by evaporation, and 25% percolates down to impermeable strata, where it is conserved at varying depths. These subterranean sources of water can be exploited, by various techniques of pumping. The underground streams and the subterranean waters drain off, in the west, towards the Mediterranean, and, in the east, into the Dead Sea, Lake Kinneret and Lake Hulah. Measurement of the flow of water in the rivers and springs presents no problems other than that of adequate time for the work of measurement. It is far more difficult to arrive at any reliable estimate of the volume of subterranean water or of floodwaters.

This is the main source of disagreement among experts in estimating exploitable water resources. Estimates commonly accepted vary between 2·4 and 3 billion cubic metres annually.

Water plans drafted in recent years are based upon the lower estimate. The moral of the anecdote related at the beginning of this chapter need not be enlarged upon. Only when all the great water projects already planned are executed and a much larger population than the present has utilized all available water, will the prospects of further sources be finally clarified.

The main sources of water are as follows:

(1) The flow of rivers and springs, varying between summer and winter and between a rainy year and one of drought.

(2) Floodwater during or after rains.

(3) Underground sources.

The aggregate of the estimated volume available annually from each of these sources is given below (in million of cubic metres):

Upper Jordan	700– 750
Yarkon	240– 250
Springs	220– 270
Underground sources	900–1,200
Floodwaters	190– 340
Purified sewage	80– 120
Underground and floodwater in the Arraba	60– 120
Return flow from irrigation	250– 350
	2,640–3,400
Losses	240– 400
	2,400–3,000 million cubic metres

In the year 1946/47 Jewish and Arab farmers irrigated a total area of 410,000 dunams within the present borders of the State.

In 1948/49	290,000 dunams were irrigated
„ 1949/50	350,000 „ „ „
„ 1950/51	440,000 „ „ „
„ 1951/52	500,000 „ „ „
„ 1952/53	580,000 „ „ „
„ 1953/54	750,000 „ „ „

The volume of water utilized rose from 300 million cubic metres in 1948 to 830 million cubic metres in 1954. It must be borne in mind, however, that in this period the country's population grew by a million new consumers—requiring an additional 100 million cubic metres annually—in addition to the various industrial plants established. Over the six years under review the volume of water consumed has increased by 177%. Nevertheless Israel, today, utilizes only one-third of its water resources, even on the basis of the most conservative estimates.

More recently water has been discovered in mountainous strata, which may increase the volume of subterranean resources. In the

Northern Negev investigations indicate the possibility of conservation in dams. In order to achieve more efficient and economical utilization various methods of using the return flow of irrigation and sewage for irrigation or industrial purposes are to be introduced.

The knowledge that in view of natural conditions in Israel, development is dependent upon intensive agriculture, based upon high and stable yields, has stimulated hydro-geological research and prospecting for water in areas hitherto not considered promising. The distribution of rainfall is such that over a distance of 600 kilometres, from the Lebanese border in the north to Elath in the south, precipitation ranges from an annual 700–800 millimetres to 50 millimetres and even less. About 60% of Israel territory is in the arid zone. It is vital for the State to settle this area, but this is possible only if water is brought here from afar. This necessity is not unique; a similar situation exists in certain parts of the United States and in Australia. The difference in regard to Israel lies rather in the fact that the wasteland covers more than half of its area.

Projects and waterworks

Three billion cubic metres of water annually suffice to irrigate 2,500,000 dunams of land, or 12% of Israel territory. Under irrigated cultivation this area could produce sufficient food for a population double the present. The distribution of the water between the various regions and the transfer of the water from the north to the central and southern districts, both constitute major problems.

The magnitude and gravity of this dual problem prompted the establishment of a water company whose operations extend over all Israel and which gained the central position in planning of the exploitation of water resources. Over the past two decades a number of district water co-operatives have been floated in the principal settlement areas, supplementing the operations of local and regional water companies. The largest of these bodies, the 'Mekoroth' Company, commenced work in 1936, in the Kishon District and in Haifa. Its activities have now become country wide.

TABLE 8

Activities of the 'Mekoroth' Company

Years	Output (cubic metres)	Irrigated area (dunams)
1938 (year)	17,814	
1948 (year)	19,820,831	24,700
1951 (31.12)	78,632,000	75,400
1952 (31.12)	117,370,000	112,000
1953 (31.12)	154,380,000	155,800
1954 (31.12)	194,260,000	225,000

The company embarked in 1946/47 upon the erection, on behalf of the settlement authorities, of the first project in the Negev.

Some years prior to the establishment of the 'Mekoroth' Company, the 'Palestine Water Company' was founded, to develop both regional and local waterworks. A notable feature of the work of this company was that the settlers did not participate in the conduct of its operations. It covered a relatively small field, and in 1951 merged with the Mekoroth Company. There are also district waterworks, mainly in the Coastal Plain, which confine themselves to supplying water to a single settlement, for the cultivation of citrus and vegetables and also for industrial purposes. This is a satisfactory and advantageous arrangement, making for economies in the use of water and obviating the expense of sinking new bores.

The Beth Shean district was notorious in the past for its obsolete and wasteful methods of utilizing water for irrigation. The Mandatory Government endeavoured to intervene to secure a more rational distribution of water among the local Arabs, but with little success. An insuperable difficulty were the claims put forward by several families, for special privileges based upon ancient usage. The Jewish settlers proved more successful in rationalizing the local water resources and in introducing an equitable system of distribution. There are fourteen dams in the district, differing in size and in the quality of their water. The water has accordingly been classified in two categories, sweet water being used for agriculture, and saline water for fishing ponds. The quota of each settlement is fixed upon the basis of the number of families, while the flow has been regulated by the construction of concrete canals.

A network of regional, local and village waterworks covers the entire area at present being irrigated but a national water project will eventually control half of the irrigated area.

This national water scheme envisages leading water from the headwaters of the Jordan on the northern frontier by way of a central conduit to the main reservoir in the Vale of Beth Netupha near Nazareth. Thence the water will flow through the Jezreel Valley, via a tunnel through the Mountains of Ephraim and then again down a conduit along the Coastal Plain to the Negev. Here feeder pipes will distribute the water to various areas. The main conduit will receive surplus water en route, to augment the supply.

Execution of this national water project will require many years. In the meantime, however, during the 1950–53 period, six short-term regional plans have been prepared. These include:

(1) The draining of Lake Hulah and irrigation of the Hulah Valley. A main conduit will divert the waters of the Danstream westwards, lengthwise across the slope, closing the valley in the south. Differ-

ences in altitude between the source of the water and the Hulah Valley will be exploited for the generation of electrical power to pump the water over the mountains on the northern border. The flow of this project will be 130 million cubic metres, of which 40 million are already being utilized. Both the drainage and irrigation schemes are at present under construction.

(2) A central reservoir to be fed by the upper headwaters of the Jordan is to be constructed in the Mountains of Galilee. This project will supply a further 30 million cubic metres. Blueprints are still in the preliminary drafting stage.

(3) Water from the lower reaches of the Jordan will be utilized for irrigation of land on the shores of Lake Kinneret, and in the Jordan and Beth Shean Valleys. This project will have a flow of 150 million cubic metres, of which 50 million are already being utilized.

(4) Surplus water from springs and wells in Western Galilee will be pumped to the Haifa municipal area, the Zebulon Valley, Central and Western Jezreel Valley, Harod Valley and the adjacent hills. Also included in this scheme are plans for the conservation of the Kishon floodwaters and the sewage effluent from Haifa. Execution of this project, which will supply 150 million cubic metres, has already begun.

(5) One hundred million cubic metres of water will be pumped through 66-inch pipes from the Yarkon to the Negev. This constitutes 40% of the flow of the river. Detailed plans for the execution of this project are being drafted and construction is already under way.

(6) A second, eastern, arm, of the same dimensions, will lead another 40% of the flow of the Yarkon to the same area in the Northern Negev. Feeder pipes will supply the Jerusalem Corridor. This project is still in the planning stage.

The aggregate flow of these supplementary projects will be 660 million cubic metres, of which 90 million are already being utilized. The present water supply will accordingly be augmented by 550 million cubic metres, enabling the irrigation of another million dunams of land.

The main feature of the national master-plan, however, will be the main conduit, a huge canal leading the water from the headwaters of the Jordan to the central storage dam at Beth Netupha, and thence under pressure through a long pipeline down to the southern part of the country.

The main reservoir in the Beth Netupha Valley will conserve one billion cubic metres, while a network of smaller dams will hold another two billion. In Israel, however, conservation of water has to contend with a major problem, as losses by percolation are high and methods of dealing with it costly, while the results are uncertain. The construction of dams will accordingly be executed in successive stages.

A scheme for the generation of electric power has been included in these water projects, the basic technique of which is to exploit the difference in altitude between the Mediterranean and the Dead Sea—which lies 390 metres below sea-level; 1,500 million kilowatt hours of power can be produced, thus saving 600,000 tons of fuel yearly.

Implementation of this project, however, will be postponed to a later stage in view of the immense investment it calls for and the international complications involved. It is assumed that the great economic benefits that can accrue will outweigh other considerations.

It is abundantly clear from the foregoing that administration of water affairs in Israel requires highly efficient organization. The water must be produced, led, stored, distributed. Economical use of water must be ensured, and large costly enterprises must be undertaken. For this reason alone it is essential that the State assume full control of all matters affecting water resources.

The sixth chapter of the Government's Programme submitted to Israel's Knesset on 11th March 1949, dealing with development plans, calls for 'the nationalization of water resources, natural resources, lands in waste areas and services vital to the security of the State'.

The administration of water resources, the cardinal element in the development of the country, is inconceivable on any other basis. The Government must appoint a Minister responsible for water affairs. His authority should comprehend control of all water resources—rivers, springs, floodwaters, subterranean sources, drainage, sewage.

The Minister for Water Affairs and the diverse organs created to assist him in his task—a National Water Council, a Water Commissioner and the like—will define water rights, in keeping with the provisions of a special Water Act which the Knesset must legislate. The main purpose of this law must be to ensure that water resources are utilized for the development of the country and principally for the benefit of agriculture, and must cover such fields as the settlement of wastelands, rational and economical use of water, supervision of investment, water tariffs in various districts, organization and administration, with the object of safeguarding the economic and social principles that underly the water problem in this country.

The Israel Government appointed a public committee to draft such a water law. After two years' work, at the beginning of 1955, the Committee submitted the draft Water Bill to the Government.

The Ministry of Agriculture established in 1948 a Water Division and appointed a National Planning Committee. The Government also appointed a Council of experts of international reputation, with extensive practical experience in the construction of large water projects in many countries, to advise it on this question.

In the beginning of 1952 the Israel Water Planning Authority was established, 52% of its capital being subscribed by the Government, 24% by the Jewish Agency and 24% by the Jewish National Fund. The function of this Authority is to plan the development of water projects, and principally the national master-plan, in keeping with the decisions of the National Planning Committee and the recommendations of the Council referred to above.

A series of cases of co-operation between the Water Planning Authority and overseas experts in which the latter gave important advice are cited below.

Musrara Brook. The harnessing of the floodwaters of the brook was suggested along the following lines: Diversion of the brook along a new course flowing into the sea, or regulation of the overflow of water by the construction of dams in the foothills of the Judean Mountains, or deepening the bed of the brook, or a combination of some or all of these solutions.

Exploitation of the Dead Sea for the generation of power. As a result of work of research the relation between the development of a hydro-electric project and the influence of Mediterranean waters on the chemical resources of the Dead Sea was established.

Main conduit. After conducting suitable research work the National Planning Committee resolved that the main water conduit, leading from the Beth Netupha Dam to the Negev, would be a pre-cast concrete pressure pipe.

Geological survey. A report prepared by an American geologist deals with the location of dams, to be integrated into the national plan situated as follows: Beth Netupha, Shuval, Gerar, Tel Yeruham, Kishon, Mishmar Ayalon, Faluja, Beersheba Brook, Yatur Brook, Mushash Brook, Nevatim, Arur Brook, Tir Brook, Muweilieh Brook, Burgot Brook, Uwein Brook and Ein Karem.

A number of examples of projects already executed or under construction follows. Plans were prepared by the Planning Authority and the work of construction has been undertaken by the 'Mekoroth' Water Company Ltd.

Yarkon–Negev pipeline. Three sections are under construction. From Magen to Cheletz—38 kilometres; a line linking up the pipes between Cheletz and the Gezir-Sumeil crossing—10 kilometres; the eastern section from the headwaters of the Yarkon to the above-mentioned crossing—53 kilometres. Pumping stations and a dam are being constructed at Tekuma. Smaller projects have been executed and are already in operation in Shuval-Erez-Lahav, in Tekuma and in Mamashit (Kurnub).

In Galilee. The construction of the underground pumping station at Malcha has been completed, three pumping stations of the mountain projects have been erected and twelve dams have been constructed in various parts of the highlands. The main Ein Dahav–Manara pipe and feeder pipes to the settlements of Misgav Am and Yiftach have been laid, as well as other pipes such as: Malcha–Dishom–Yiftach; Dishom–Gush Chalav; Sasa–Meron; Meron–Ein Zeitim–Shamau; A Naba–Tarschicha–Hossen–Alkosh.

In the Beth Shean district the laying of the north Jusak pipe, leading water to Sdeh Nachum, and the south Jusak pipe, to supply Reshafim, Shluchot, Ein Hanaziv and the Regional Experimental Station, is rapidly approaching completion. The construction of the Fuar and Madua canals leading to the upper junction, and the south-western canal leading to the town of Beth Shean and the upper water-mixing station will also shortly be completed.

Samaria and the Arab triangle border. The Yavetz, Beerotaim and eastern pipe have been laid and put into commission. The main southern reservoir has been constructed and seven wells have been sunk.

In the Southern District and South Judea. The flow of the Lydda Plain project has been increased by sinking many new wells and the construction of additional feeder pipes. A number of reservoirs have been constructed, including one near Ramleh with 20,000 cubic metres capacity. The laying of a second pipeline to Jerusalem and the eastern section of the Corridor has been completed. The Gath waterworks which supplies the area north of the Migdal–Faluja road has been completed.

RESERVOIRS

The construction of reservoirs and lakes for the conservation of water is an engineering innovation in Israel. Conditions for the construction of such reservoirs are extremely unfavourable and the collaboration of experts, with experience in this type of work in other countries, has been sought. The construction of reservoirs enables the conservation of water from the winter to the summer, and from a year of abundant rains to one of drought. In the absence of such dams the volume of water that can be utilized will be halved.

The main problem encountered in the construction of these reservoirs is the large volume of water lost through seepage down the slopes and leakage through the foundations of the dam. More exact knowledge of the behaviour of materials under conditions of flooding, the degree of seepage, methods of surfacing and the creation of a link with the bed rock is still required. For this reason extensive research has been conducted over a lengthy period in the Beth Netupha Valley where it was found that the large quantity of naked rock would cause a high degree of seepage and percolation. Effective and inexpensive methods of preventing loss of water in this large dam were tested. It was also found that the clay soils of Galilee and the loess soils of the Negev are suitable for the erection of stable dams. In Galilee the location of a number of reservoirs is of porous basalt, with many clefts and fissures, which will have to be dealt with specially. In practically all locations a deep intervening wall in the foundation of the dam and cementing is necessary to prevent leakage under the dams. Where it is impossible to prevent leakage the water will be allowed to penetrate down to an impermeable stratum, to be recovered later by pumping in the same way as subterranean water.

Experimental dams have been constructed in Shuval, Beersheba Brook, Gerar Brook and Tel Yeruham. Various phases and methods of experimentation have been planned before final conclusions are drawn.

In addition to these experimental dams, larger reservoirs have been constructed, including the Beth Netupha, Mishmar Ayalon, Kishon and Ein Karem Dams. We shall confine ourselves to a survey of the most important of all of these, namely, the Beth Netupha Dam.

The main function of the Beth Netupha Dam, which is to serve as the central reservoir for a billion cubic metres of water, is to justify the large investment required for grouting. It will be necessary to create an impermeable protective stratum, particularly in the eastern section of the dam. The collection basin of the dam extends over an area of 90 square kilometres, of which 68 square kilometres are naked rock and the remainder dirt. Local rainfall will enable the conservation of several million cubic metres of water, sufficient for experimental purposes. A dam wall has been erected across the Haldia Brook, 450 metres in length, 13 metres wide and 5 metres thick along its upper edge. The total volume of this dam wall is 133,000 cubic metres. The capacity of the dam at the highest water-level is 12 million cubic metres. Its purpose for the present is to measure the percolation coefficient in the Valley, along the slopes and on the bare rock, to test methods of construction, grouting and surfacing, and the stability of foundations of the dam. During the winter of 1952/53, 1·5 million cubic metres of rainwater were stored in the dam to a height of 142·8 metres, without any substantial water losses. In the following winter 6 million cubic metres were conserved with little loss of water. It is proposed to allow the dam to remain in this state for a year or two to enable more prolonged tests before a final decision is taken regarding the possibility of storing scores and even hundreds of millions of cubic metres of water in it.

THE ARRABA

The Engineering Corps of the Israel Defence Army pioneered a regular supply of water to Elath, by the organization of a tanker service and by the purification of saline water on the spot. These methods permitted, however, only a limited supply. In view of the increasing demand for water and initial attempts to develop Elath and its immediate neighbourhood, the Water Division of the Ministry of Agriculture began to sink wells near Elath and Sabcha, in the hope of striking subterranean water. The water struck by eighteen test borings was saline in every case. Good water with a chlorine content of 250–600 milligrams was discovered at Beer Ora (Bir Hindes) 18 kilometres from Elath, but the flow was only moderate. The Engineering Corps laid down a six-inch pipe from Beer Ora to Elath, where a 200 cubic metres storage reservoir and a system of distributing pipes was constructed. The Water Division constructed two reservoirs, each of 1,000 cubic metres capacity at Elath and Beer Ora respectively, to ensure a regular supply of water, as the flow of the boreholes was erratic. Twelve other bores in the same vicinity did not result in any improvement in the supply. Only a bore at Ein Radian, 40 kilometres north of Elath, was successful, producing 260 cubic metres of drinking water per hour. At Ein Yahav, 120 kilometres north of Elath, the flow of three local springs, with a combined flow of 25 cubic metres per hour, was conserved to serve the needs of the local experimental farm.

Ten bores were sunk in the Paran Brook (Giraffi). Six of these were unsuccessful, while the remaining four produced small quantities of water. However the fact that this watercourse constitutes a channel for a large volume of floodwater has given rise to a hypothesis that by the damming of the water and the construction of an artificial lake, the flow will be enabled to percolate down to a stratum of rock bearing subterranean water, which it will prove possible to exploit by artesian boring. These projects, however, are still in the early stages of planning and research.

DRAINAGE

Paradoxically enough, in addition to the normal problems encountered in an arid country like Israel, the solution of which is both costly and difficult, there is also the problem created in certain areas by surplus water. Water surpluses may cause harm by giving rise to floods, swamps, a high subterranean water-table and impermeable soils.

The watercourses in the plains and particularly in the Coastal Plain are incapable of draining off the winter rainfall which flows down in sudden torrents from the mountains, flooding fields and residential areas. In some areas where there is no outlet for the rainwater into the sea, marshes have been formed.

A high level of subterranean water may be caused by the existence of impermeable strata; the resultant excessive moisture in the soil is harmful to agriculture. This is the case in the Northern Hulah Plain, in Kabbara and certain areas in the Jordan Valley.

Where the water-table reaches a level higher than the surface, marshes like the Naaman swamp are formed.

On heavy soils, rainwater and water from irrigation remain in the upper layer of soil, blocking the vents, preventing ventilation and causing the roots of the crops to rot or degenerate before the plant has had a chance to develop.

Surplus water can prove as harmful as lack of water, though its effects may assume different forms. Damage to crops, epidemics of fever, stoppages of traffic and the isolation of villages, which can affect their security, are some of the consequences of floods.

The most effective preventive measure is drainage.

The projects which have engaged the attention of the water authorities are the drainage of the Hulah and Gaaton Valleys, the Naaman District, the Falik Brook, the Nesher District, the Kishon Valley and the Malik Brook.

Research is being conducted into the drainage needs of the country.

The most important scheme in hand at present is the draining of the Hulah Valley.

Some 57,000 dunams of land in the Hulah Valley were classified as 'Jiftlik'[1] under the Ottoman regime. It was the practice of the Turkish Government to transfer the more important of the Jiftlik lands to concessionaires, either individuals or public companies. A concession over the Hulah Valley was granted to two Arab families for approximately £5,000, conditional upon the concessionaires undertaking to drain the marshes within a specified period. Under the terms of the concession 10,000 dunams of the drained land was to be granted to Arab cultivators in the neighbourhood. The title of

[1] 'Jiftlik' is a Turkish term, meaning 'double'. When used in relation to the soil it defines an area cultivable with a span of oxen. In Palestine the term was used to describe land which had reverted to the Ottoman Government for various reasons, and had become the property of the Sultan.

the concessionaires to the area was to be recognized only after fulfilment of these conditions. In 1914 the concession was transferred to the Syrian Ottoman Agricultural Company, which in 1924 obtained provisional approval of the concession from the Mandatory Government. In 1929 this approval was made absolute, the term for the completion of the drainage works being extended to 1937. The concessionaires were unable to fulfil the conditions of the grant and consented to dispose of their rights to a Jewish company and in 1934 the concession was registered in the name of the Palestine Land Development Company. Hulah Valley and the area north of the Valley were acquired subsequently and settlement work was launched in 1939. Concurrently preliminary investigation for the planning of the drainage project was conducted, mainly by British engineers.

A committee of experts appointed by the Government of Israel recommended shelving the proposals of the British engineers, and decided to drain the floodwaters by excavating canals transversing the low-lying areas in the eastern and western sections of the valley. The plan contemplated complete drainage of the marshes with the exception of the peat deposits, and lowering the water-level in the adjacent areas. A system of canals will drain the floodwaters from the valley in a short time. The peat deposits will be kept saturated with water by the construction of a ramp, to facilitate exploitation of the peat. This area will at the same time serve as a reserve of wild life and flora.

For a period of three years extensive excavation works, involving the deepening of the course of the Jordan to lower the level of the water in the lake, have been carried out. A system of dams makes it possible to keep the water in the lake at any level desired.

Smaller drainage works were conducted in seven other districts in these years.

ANTI-EROSION MEASURES

Only the upper strata of the soil are fertile and can produce crops. Mostly the layer of soil sustaining plant life is superficial, and is underlaid by less fertile strata, including rock, sand or clay. Even on deeper soils fertility ceases at a moderate depth, for biological life no longer exists. For this reason it is incumbent upon Man to preserve the top soil from destruction, erosion and impoverishment. When the upper layer of the soil lacks a protective covering of vegetation it is most vulnerable to the agents of destruction. It is only in the past two decades that soil conservation has been professionally developed and technical measures for the prevention of erosion have been tested.

The soil of this country, which hundreds of years ago was fertile and highly productive, has been ruined as a result of neglect and the effects of rain, wind, the depredations of goats, grass fires and backward methods of cultivation. Shortly before the establishment of Israel local cultivators began to appreciate that adequate measures must be adopted to conserve the soil. Such a task, however, is beyond the means or capacity of the individual farmer and the aid of the State is required in planning, research, legislation and finance.

A Division of Soil Conservation was created under the aegis of the Ministry of Agriculture for the implementation of various measures for soil conservation. Its main task, however, has been the prevention of erosion. A station was established to study erosion and water-flow problems in the coastal district.

The functions of the station have been defined as follows: (a) to gather data on precipitation, upper water flow and erosion under various conditions of cultivation of the soil; (b) to test the efficacy of various soil conservation techniques; (c) to demonstrate the utility of conservation methods in diverse settlements.

Conservation work has been launched in dozens of settlements with a view to determining the flow of water, testing construction of protective canals, stoppage of fissures, cultivation of pasture grasses, rehabilitation of existing shrubs and the planting of new shrubs, terracing, planting of wind breaks, contour de-stoning, harnessing of floodwaters, the construction of small dams, prevention of forest-fires, sowing and planting along highways to prevent erosion. Drainage canals have been excavated, woods and plantations have been planted.

4

CHANGES IN THE ARAB VILLAGE

PRIOR to the establishment of the State of Israel, there were, within its present boundaries, 364 Arab villages and 10 towns, excluding the Beersheba District. Of the total area of 4,311,000 dunams covered by these towns and villages, 2,750,000 dunams of village land and 180,000 dunams of urban land (out of a total of 263,000 dunams) were under cultivation. Bedouin tribes in the Beersheba District cultivated two million dunams out of a total of 12·5 million dunams.

The distribution of these lands in the various districts, excluding the Beersheba District, was roughly as follows:[1]

TABLE 9

Arab villages—number and area in two periods

District	Prior to May 1948			Since May 1948		
	No. of villages	Total area (dunams)	Cultivated area (dunams)	No. of villages	Total area (dunams)	Cultivated area (dunams)
Southern	149	1,632,000	1,217,000	7	10,000	10,000
Central	27	333,536	256,000	13	145,000	90,000
Haifa	36	483,000	279,000	9	221,000	62,000
Galilee	152	1,862,000	1,000,000	59	860,000	391,000
Total	364	4,310,536	2,752,000	88	1,236,000	553,000

With the exception of the inhabitants of three villages in the Jerusalem Sub-district, virtually all Arab residents of the Southern District have left the country. Small areas are cultivated by Arabs resident in Lydda, Ramleh and two villages in the Tel-e-Safi area (near Kfar Menachem).

[1] *Village Statistics*, Government of Palestine, Jerusalem, 1 April 1945.

In the 'Triangle', however, the inhabitants of about half of the villages have remained. In the Carmel District one-quarter of the villagers have remained, cultivating one-half of the former area. Inhabitants of the Safad, Beth Shean and Tiberias Districts, for the most part, have left their villages. As against this the majority of the villagers in the Nazareth and Acre Districts have remained.

In April 1945 the population of these 364 villages (excluding Beersheba) was 372,000. The average population per village accordingly was approximately 1,000 souls and the area 12,000 dunams, of which 7,500 dunams were under cultivation.

The average population and area of the villages varied from district to district, as will be seen from the following table:

TABLE 10

Average Arab rural population and areas in 364 villages (1945)

District	Average population per village	Average area per village (dunams)	Average cultivated area per village (dunams)	Average area of orchards per village (dunams)	Area per capita (dunams)	Cultivated area per capita (dunams)
Southern	1,204	10,956	8,165	1,348	9·1	6·8
Triangle	1,054	12,353	9,483	838	11·7	9·0
Haifa Sub-district	1,114	13,425	7,739	628	12·0	6·9
Galilee	816	12,248	6,542	1,239	15·0	8·0
Whole country	1,022	11,844	7,543	1,193	11·6	7·4

These averages were highly representative of the situation in the Arab villages during the Mandatory administration.

The 88 Arab villages, which have remained in the State of Israel, had a total population of 96,000 inhabitants in 1945. Each village accordingly averaged 1,090 inhabitants upon an area of 16,450 dunams, of which 9,100 dunams—including 1,400 dunams of orchards—were under cultivation.

The average population and area of the Arab villages in the various parts of the country in 1945 were as indicated in Table 11 on the following page.

According to a census conducted by the Central Bureau for Statistics and Economic Research on 31st December, 1951, there were 104,500 inhabitants in the Arab villages in the State of Israel, occupying 553,000 dunams of land, of which 80,000 dunams were abandoned lands which had been leased to Arab cultivators by the Custodian of Absentee Property.

TABLE 11

Average Arab population and areas in 88 villages (1945)

District	Inhabit-ants	Average area per village (dunams)	Average cultivated area per village (dunams)	Average area of orchards per village (dunams)	Area per capita (dunams)	Cultivated area per capita (dunams)
Southern	798	7,062	4,984	1,029	8·8	6·2
Triangle	1,714	19,490	14,490	1,467	11·4	8·4
Haifa Sub-district	1,546	21,284	12,316	1,050	13·8	8·0
Galilee	912	15,919	7,866	1,453	17·4	8·6
Whole country	1,089	16,492	9,136	1,388	15·1	8·4

Each of these villages averaged 1,250 inhabitants occupying 14,700 dunams, of which 6,600 were under cultivation.

The following table reflects the situation in the various parts of the country. All figures are averages.

TABLE 12

Average Arab rural population and areas in 1951

District	Popula-tion	Total area (dunams)	Cultivated area (dunams)	Orchards area (dunams)	Total area per capita (dunams)	Cultivated area per capita (dunams)
South	453	3,000	3,000	274	7·4	7·4
Triangle	2,113	11,125	7,044	893	5·3	3·3
Haifa Sub-district	1,327	24,530	6,889	536	18·5	5·2
Galilee	1,080	14,576	6,627	968	13·5	6·1
All Israel	1,244	14,707	6,602	885	11·8	5·3

Comparison of the figures for the Arab villages in 1951 with those for the Mandatory period, indicates a rise in the average population and a drop in both the total area occupied and the area cultivated. In 1951, the fixed populations of the villages—in addition to 20,000 Arab refugees—occupied a cultivated area that was 150,000 dunams smaller than that cultivated by these villages during the Mandatory regime.

SYSTEMS OF LAND TENURE

In 1930 the Government of Palestine conducted a survey[1] com-prehending 104 Arab villages in various parts of the country. In

[1] By the Johnson-Crosbie Commission.

these villages 68·2% of lands was cultivated by the proprietors and the remaining 31·8% by tenants. The census conducted by the Government of Israel in 1950 revealed that 78·5% of the land was privately owned and 21·5% held on leasehold. 95·6% of the farms were privately owned, while 30·5% of the farms were on land partly privately owned and partly leased. Only 4·4% of the farms occupied leased land. These statistics refer only to farms five dunams and more in extent.

No significant changes have been registered in the conditions of tenancy. In general rent is paid in kind, in the form of a percentage of the crop—up to one-third—though there are cases of cash rentals.

The proportion of land under cultivation in private ownership has increased somewhat since the termination of the Mandatory regime, at the expense of the land cultivated by tenants. This may be explained by the fact that many of the larger landlords left the country during the War of Liberation. Their lands were taken over by the Custodian of Absentee Property and are no longer cultivated by Arab peasants.

Another interesting feature is the disparity in the size of Arab farm-units. A survey of five villages made in 1944, in the Lydda and Ramleh Sub-districts, revealed the following pattern of Arab land ownership.[1]

TABLE 13

Size of Arab farms in 1944

Size of farm (dunams)	No. of farms	% of all farms	Total area (dunams)	% of total area (dunams)
Less than 20	121	30·0	948	3·5
21–60	118	29·3	4,605	17·0
61–80	43	10·7	2,948	10·9
81–120	52	12·9	5,235	19·3
More than 120	69	17·1	13,380	49·3
Total	403	100·0	27,116	100·0

The average area of these farms was 67·3 dunams, but 60% of the total were less than 60 dunams in extent.

The survey conducted in 1950 produced a different picture, reflected in Table 14 overleaf.[2]

[1] *A Survey of Palestine*, prepared in December 1945 and January 1946 for the information of the Anglo-American Committee of Inquiry, Government of Palestine, Vol. I, 1946, p. 277.

[2] Based upon Table 3, pp. 24–25. *Agricultural Census*, 1950, Part I, issued by the Central Bureau of Statistics and Economic Research, Government of Israel.

TABLE 14

Size of Arab farms in 1950

Area of farm (dunams)	No. of farms	% of all farms	Area of farms (dunams)	% of total area (dunams)
Less than 19	5,343	44·5	43,509	8·1
20–49	3,262	27·2	104,877	19·7
50–74	1,299	10·8	78,825	14·8
75–99	760	6·3	65,207	12·2
More than 100	1,346	11·2	241,433	45·2
Total	12,010	100·0	533,851	100·0

Here the area of the average holding (for the whole country) was 44·4 dunams, with approximately 60% of the farms occupying less than 30 dunams. The area of the land-unit in Arab agriculture declined, as will be seen from the above, as a result of the decrease in the number of large and medium-sized farms.

Attention was drawn in our remarks on the survey of the five villages in Lydda–Ramleh area, to the fact that some 60% of the total number of farms cultivated an area of less than 60 dunams each. The 1950 Census revealed that the number of farms cultivating less than 50 dunams was 72% of the total. These small farms now occupy 27·8% of the total area, while in 1944 farms less than 60 dunams in extent occupied 20·5% of the total area. Further, the survey of five villages showed that large and medium-sized farms, cultivating more than 81 dunams, constituted 30% of the total number of farms, and cultivated 68·6% of the total area. Comparable figures of the 1950 Census (for farms cultivating more than 75 dunams) showed that the large and medium farms comprised 17·5% of the total number of farms and 57·4% of the area cultivated. Thus the increase in the number of small farms was accompanied by a decline in the number of large and medium farms.

A negative feature of the Arab system of land tenure is the fragmentation of the holdings. Table 15 on opposite page shows the degree of fragmentation in the five villages covered by the survey of 1944.[1]

The number of parcels in each farm is almost directly proportionate to its size. In 1944 each farm comprised an average of 10·3 parcels, each 6·5 dunams in extent. Fragmentation of Arab farms in 1950 is reflected in Table 16 on opposite page.[2]

[1] *A Survey of Palestine*, Vol. I 1946, p. 377.
[2] *Agricultural Census*, Part I, pp. 24–25.

TABLE 15

Degree of fragmentation of land-units in Arab villages (1944)

Size of farm (dunams)	No. of farms	No. of holdings	% of total	Area (dunams)	Average size of holding (dunams)	No. of holdings per farm
Less than 5	50	106	2·5	115	1·1	2·1
6–10	33	115	2·8	270	2·3	3·5
11–20	38	173	4·2	563	3·25	4·6
21–40	68	567	13·6	2,122	3·7	8·3
41–60	50	467	11·2	2,483	5·3	9·3
61–80	43	585	14·0	2,948	5·0	13·6
81–120	52	748	18·0	5,235	7·0	14·4
121 and more	69	1,402	33·7	13,380	9·5	20·3
Total	403	4,163	100·0	27,116	6·5	10·3

TABLE 16

Degree of fragmentation of land-units in Arab villages in 1950

Area of farm (dunams)	No. of farms	No. of parcels	% of total number	Area (dunams)	Average area of parcel (dunams)	Average number of parcels per farm
1–4	1,781	1,210	1·7	4,423	3·7	0·7
5–9	1,512	3,666	5·1	10,359	2·8	2·4
10–19	2,050	7,658	10·7	28,727	3·7	3·7
20–39	2,439	14,448	20·2	68,743	4·8	5·9
40–99	2,882	26,010	36·5	180,166	6·9	9·0
100–149	753	8,973	12·6	89,386	10·0	11·9
150 and more	593	9,427	13·2	152,047	16·1	15·9
Total	12,010	71,392	100·0	533,851	7·5	5·9

A pattern similar to that noted in the above-mentioned survey of the five villages, in respect of fragmentation of farms, still exists, the number of parcels increasing in almost direct proportion to the size of the farm, though in the present instance a definite improvement is apparent. The average number of parcels to the farm has decreased to 5·9, while the average size of each parcel has grown to 7·5 dunams. Thus while the curse of fragmentation still rests upon Arab agriculture, a trend towards improvement can be discerned.

ARAB FARM ECONOMY

The Arab farm economy has always been and still is based upon the one-family, self-sufficient unit. The changing mood of the market has exercised little influence upon Arab agriculture, though in recent years its impact is becoming more manifest. So far, unlike the Jewish sector, changes in the Arab farm economy have been very restricted in effect. The structure of Arab farming is broadly based upon the following three branches:

Grains	approximately	40%
Vegetables, fruit and olives		40%
Livestock		20%

Methods of cultivation on the farmer-owner unit, as on that of the tenant farmer—and indeed no less on the large farm—are still primitive. Implements are of the simplest, the plough hardly more than scratching the soil, manuring is not practised, and huge manure heaps, augmented by each successive generation, is a common sight in every Arab village. Indeed, it was only following the development of Jewish agriculture that manure became a marketable commodity, which, as a result of the constant demand by Jewish citrus, fruit and vegetable growers, gradually disappeared in many areas over the past half-century. In sharp contrast to Jewish farming, the various branches on the Arab farm—though based also on mixed agriculture—are not closely integrated. In the majority of cases one cash crop, such as tobacco in Galilee, olives or livestock in the highlands, field crops in the plains, is the principal or even only source of income. The main purpose of whatever other branches of farming may be pursued is to meet domestic consumption.

The Arab peasant, despite his primitive way of life and methods of cultivation, is a true son of the soil, living off his land. He is unmoved by the vagaries of the external economy. In a good year he lives well and even puts something by for a rainy day. When lean times come perforce he tightens his belt. He lives within the limits of his means.

PRODUCTION

During the Second World War Arab farming experienced a period of comparative prosperity, as a result of the virtual stoppage of food imports. Local farm produce, with the exception of citrus, was easily absorbed at high prices by the local market. This state of affairs even stimulated the introduction of new crops such as potatoes, ground-nuts and certain varieties of vegetables.

In physical dimensions Arab agriculture by far exceeded that of the Jewish sector prior to the War of Liberation. Arab produce to a large extent competed with the produce of the Jewish farmers, although, because of better methods of marketing, publicity and campaigns in favour of Jewish produce, the latter were able to hold their ground.

In the year 1944/45 Arab farms comprised 93% of the area under grains, vegetables, fruit (excluding citrus), fodder, olives and melons. Jewish farms accounted for the remaining 7%. In production, however, the situation was different. Arab produce constituted 71% of the total and Jewish produce—29%.

The distribution (in 1944/45) of the area under the crops referred to, as well as the value of production, is given in the following table:

TABLE 17

Arab and Jewish areas, crops and returns in 1944/45

	Area (dunams)	Field and garden crops %	Fruit (excluding citrus)	Returns £I.
Arabs	5,500,000	83	17	22,000,000
Jews	425,000	90	10	4,700,000

In regard to area under cultivation the ratio was 13 : 1 in favour of the Arabs, and for value of production 4·7 : 1. In that year field crops accounted for 62% and fruit for 38% of the total cash income of the Arab farmers. For the Jewish farms the figures were 90% and 10% respectively.

Prior to the establishment of the State of Israel, Arab farms occupied an area of 4,574,000 dunams in extent (excluding the Negev), of which 2,930,000 dunams were under cultivation. The Arab sector of Israeli farming today comprises 1·25 million dunams, of which 500,000 dunams are being cultivated. In addition Bedouins cultivate some 200,000 dunams in the Negev.

The following table shows the area cultivated by Arabs during 1945–51.[1]

TABLE 18

Areas under field crops and orchards in Arab villages in 1945–51

Year	Field crops (dunams)	%	Orchards (dunams)	%	Total area (dunams)
1 April, '45	2,444,883	83·6	479,034	16·4	2,923,917*
1948/49	441,400	85·8	73,000	14·2	514,400
1949/50	484,000	86·9	73,000	13·1	557,000
Census 1950	409,196	84·1	77,193	15·9	486,389
1950/51	496,800	86·7	76,000	13·3	572,800

* The area of 1945—in Palestine; since 1948/49—in Israel.

[1] (a) *Village Statistics.* (b) *Periodical Reviews of the Arab Farming Division,* Israel Ministry of Agriculture. (c) *Agricultural Census,* 1950.

There has been little change, it will be noted, in the relative area under field crops and orchards respectively, during the period reviewed.

The value of Arab farm production has grown steadily in these years (totals are based upon 1948/49 prices):

> 1948/49—£I.2·8 million
> 1949/50—£I.4·75 „
> 1950/51—£I.3·6 „ (a year of drought)
> 1951/52—£I.5·9 „

In the course of these four years the value of farm production has more than doubled. According to the price levels reigning each year the value of Arab production was as follows:

> 1948/49—£I.2·8 million
> 1949/50—£I.5·1 „
> 1950/51—£I.5·6 „
> 1951/52—£I.17·1 „

Computed in terms of 1948/49 prices the value of Arab farm production fluctuated between 7·5% and 9·4% of the gross farm production in Israel. Upon the basis of price levels reigning each year it ranged between 7·5% and 12·3% of the total.

The drought of 1950/51 struck Arab farmers harder than it did the Jews, as their more extensive methods of agriculture made them more dependent upon natural conditions.

Field crops, including tobacco, constituted the main category of crops cultivated in 1948/49. These were followed by livestock, orchards and vegetables (in that order). The proportion of these categories in the Arab farm economy (of gross value) was: field crops—35%; livestock produce—22%; fruit—16%; vegetables—15%.

It is interesting to note that the relative importance of tobacco cultivation has risen fourfold in these four years.

The following table on opposite page reflects comparative changes, in area, production, prices and value of yields of the main crops over the years 1949–52 (base: 1948/49 = 100).[1]

The table indicates a correlation between the low prices for wheat and barley in 1949/50 and 1950/51 and a decline in the area sown with these crops in the two years that followed. With regard to maize and durrha, however, a contrary development can be noted, for despite the fact that prices for both were low in 1949/50 in relation to the general price level for the previous year the area under maize and durrha increased in 1950/51. In that year again prices

[1] *Bulletins of Central Bureau of Statistics and Economic Research*, Government of Israel.

TABLE 19

Changes in area, yield and prices of field crops

Crop	1949/50			1950/51			1951/52		
	Area (dunams)	Yield (tons)	Price level	Area (dunams)	Yield (tons)	Price level	Area (dunams)	Yield (tons)	Price level
Wheat	213·3	155·3	87·5	208·2	74·5	125·0	183·0	187·2	229·2
Barley	682·6	276·7	84·4	564·1	83·3	148·4	409·0	273·3	269·4
Maize and durrha	73·1	88·9	105·6	77·8	18·5	175·0	62·1	63·0	356·2
Melons and pumpkins	196·6	560·0	265·4	128·5	103·6	395·4	207·5	276·4	1026·4
Tobacco	306·0	241·7	132·3	514·6	300·0	242·9	604·4	408·3	316·9
Potatoes	100·0	133·3	76·1	154·6	83·3	75·1	38·0	50·0	200·0
Sundry Vegetables	100·0	275·3	106·5	85·6	236·0	161·0	69·6	255·1	364·0
Average price level of all farm produce*	—	—	115·1	—	—	166·3	—	—	299·6

* The formula used for calculating the index of the general price level for Arab farm produce is

$$I = 100\sqrt{\frac{\epsilon p_i q_0}{\epsilon p_0 q_0} \cdot \frac{\epsilon p_i q_i}{\epsilon p_0 q_i}}$$

p_0 = each price per unit in the basic year (1948/49).
q_0 = the quantity of produce in the basic year.
p_i = price per unit of produce in the year under review as compared with the basic year.
q_i = production in year under review.

We have not included the magnitude or value of natural increase of livestock remaining on the farms. The index accordingly is based upon produce consumed on the farm or marketed.

were above the general level, but the area under both maize and durrha declined in the following year. In regard to melons, in every year prices obtained were higher than the general price level. Nevertheless the area sown with melons was smaller in 1950/51, though it increased again in the following year.

Tobacco cultivation provides an instructive example of the close relation between rising prices and expanding cultivation, as may be expected in the case of an industrial crop. The price index for tobacco was above the general level in each year, while the area planted grew steadily. Tobacco growing in this country is mainly confined to the Arab sector.

The price index for vegetables was lower than the general index for agricultural produce, but rose rapidly after the removal of Government controls in 1952. Vegetable growing contracted in 1951 and 1952 as a result of comparatively low prices.

The brevity of the period reviewed and the gaps in the data given do not justify any conclusions as to the extent relative prices obtained for various crops influence the agricultural planning of the Arab farmer. Broadly we may venture an opinion that marketing considerations are not decisive in this respect, as Arab farming is still based upon production for domestic consumption. At the same time, it may be assumed that demand and market conditions do influence the planning of production in respect of cash crops, such as vegetables, cucurbits and tobacco.

The Ministry of Agriculture has encouraged the Arabs to extend the area under irrigated cultivation, which indeed grew from 2,500 dunams in 1948/49 to 7,400 dunams in 1950/51. This increase, however, is minimal in relation to the area cultivated by the Arabs. Most of the fields crops, vegetables and orchards are not irrigated and are therefore particularly at the mercy of weather conditions. This, and not changes in the area under the various crops, is the main reason for fluctuations—in some years very sharp fluctuations —in the yield. The sensitivity of Arab agriculture to climatic conditions is demonstrated by the vast difference in yields between the drought-stricken year of 1951 and that of 1952, which was marked by copious rains and good weather. The following table provides comparative data for production planning and the influence exerted by various factors upon the yield, for the years 1950/51 and 1951/52.

TABLE 20

Indices of changes in area and yield of field crops

Crop	1950/51 (1949/50 = 100)		1951/52 (1950/51 = 100)*	
	Area	Yield	Area	Yield
Wheat	97·6	48·0	87·9	251·3
Barley	82·6	30·1	72·5	328·1
Maize and durrha	106·4	20·8	79·8	340·5
Cucurbits	65·4·	18·5	161·5	266·8
Tobacco	168·2	124·1	117·4	136·1
Potatoes	154·6	62·5	24·6	60·0
Vegetables	85·6	85·7	81·3	108·1

* Based on the previous table.

The above table indicates that in the two years selected as representing exceptional climatic conditions—drought and copious rainfall respectively—the effects of planning were less important than those of other factors.

The following table on opposite page gives the value of production of field crops, vegetables and fruit of Arab cultivators, for the years 1945–52, according to 1949 and current price levels.[1]

[1] (a) For the year 1945—A Survey of Palestine, Vol. I, 1946, p. 326. (b) For the years 1949–52: Bulletins of Central Bureau of Statistics and Economic Research, Government of Israel.

TABLE 21

Value of production of field crops

Year	Total income (£I.)	Field crops (£I.)	Tobacco (£I.)	Vegetables (£I.)
1945	17,103,133	5,529,886		5,113,553
1949	1,960,500	791,400	210,000	304,600
1950*	3,468,200	1,619,300	507,500	859,400
1951*	2,187,000	448,250	630,000	709,500
1952*	4,027,600	1,253,100	875,500	759,000
1950†	3,940,300	1,800,700	671,000	855,000
1951†	4,121,700	774,000	1,530,000	1,094,200
1952†	13,712,100	4,972,400	2,717,000	2,664,000

Year	Fruit (£I.)	% Field crops	% Tobacco	% Vegetables	% Fruit
1945	6,459,694	32·3		29·9	37·8
1949	654,000	40·4	10·7	15·5	33·4
1950*	482,000	46·7	14·6	24·8	13·9
1951*	399,250	20·5	28·8	32·4	18·3
1952*	1,140,000	31·1	21·7	18·9	28·3
1950†	613,100	45·7	17·0	21·7	15·6
1951†	723,500	18·8	37·1	26·5	17·6
1952†	3,358,700	36·3	19·8	19·4	24·5

* 1948/49 prices.
† Current prices. In 1945 in Palestine pounds, in other years in Israeli pounds.

FIELD CROPS, ROTATION, YIELDS

Field crops continue to constitute the central pillar of Arab farming, as they did under the Mandatory regime. Arab agriculture is based upon the two-field system, half of the land being sown with winter crops—wheat and barley, and half with summer crops—durrha, sesame and, in some parts of the country, watermelons. Generally speaking the land is not permitted to lie fallow. Ploughing is superficial, crop rotation simple and, in view of the absence, over many generations, of any form of manuring, yields are very meagre.

The Arab peasant still sows his field with wheat and barley, despite the low price they command in the market, both because of his innate conservatism and his primary concern to provide food for his household.

In respect of summer crops, however, a considerable change has been registered, and the cultivation of durrha and sesame, which in the past constituted the main crops, has declined while tobacco, unirrigated vegetables, cucurbits, all of which are far more lucrative, are being cultivated to an increasing extent.

Since the foundation of the State the Arabs have begun to use tractors to plough their lands in the plains, the use of chemical fertilizers is also spreading. In the course of a few years the yield of wheat has risen to 70 kilograms to the dunam, of barley to 100 kilograms, with a record yield of 180 kilograms. Not very long ago the average yield per dunam was 40–50 kilograms for wheat and 50 kilograms for barley.

The table given below shows the average yields per dunam of the main crops cultivated by Arab farmers during the years 1948/49–1951/52, and the comparative figures for the 1935–41 period.

TABLE 22

Comparative average yields of main crop (Kgs.)

Crop	Average 1935–41	Average 1948/49	Average 1949/50	Average 1950/51	Average 1951/52
Wheat	44	55	40	20	57
Barley	35	87	35	13	58
Maize and durrha	54	71	86	17	70
Legumes	—	52	26	13	47
Melons	—	274	782	221	498

Note: 1950/51 was a year of drought.

ORCHARDS

Olives, vines and figs comprised 90·7% of the total area under orchards in Arab farms in 1948. In 1950, according to the Census of Agriculture, this figure was 93·4%. In this period little change was registered in the distribution of orchards in the Arab sector, but in recent years Arab interest in the olive has waned, and new planting comprises mainly deciduous fruits and vineyards.

The following table shows production trends and the value of production of fruit in the Arab sector of agriculture, during 1949–52.[1]

TABLE 23

Yields and value of fruit in Arab farms

Variety	Unit	1948/49	1949/50	1950/51	1951/52	Index 1952 1948/49 = 100
Grapes (table)	tons	1,600	1,700	1,500	2,700	169
Olives	,,	6,900	2,400	1,750	11,450	166
Bananas	,,	—	10	20	20	200
Deciduous fruit	,,	500	800	250	700	140
Miscellaneous	,,	1,300	3,000	3,050	3,000	231
Total value of produce*	£I.	654,500	482,000	399,250	1,140,000	174
Total value of produce †	£I.	654,500	613,100	723,500	3,358,700	513

* 1948/49 prices.   † Current prices.

[1] Based upon figures published by the Central Bureau of Statistics and Economic Research, Government of Israel.

After field crops, fruit-growing is the most important branch of Arab farming. There was a considerable relative price increase for fruit mainly in 1951/52, which provided an incentive to Arab cultivators to extend their orchards.

LIVESTOCK

According to an estimate made in 1943 livestock on Arab farms numbered one million head, of which cattle accounted for 23%, sheep—58%, donkeys—11%, camels—4%, horses and mules—3% and pigs—1%.

In 1945 the Arab villages, today within the borders of Israel, owned 116,000 head of cattle, sheep, etc., the composition remaining more or less as given above. The Agricultural Census of 1950 revealed that the Arabs owned 124,000 head of livestock of which 15% were cattle, sheep—66%, donkeys—11%, horses and mules— 3% and camels—4%. Thus the distribution of livestock has remained approximately constant. The Arab peasant keeps low-grade animals of local breeds, under conditions of natural grazing eked out by a little fodder in the winter months.

The following table shows the development of the various branches of livestock farming during the years 1949-52, measured in terms of produce.[1]

TABLE 24

Livestock produce in Arab farms

Produce	Unit	Yield 1948/49	Yield 1949/50	Yield 1950/51	Yield 1951/52	Index 1952 1948/49=100
Cow's milk	1,000 litres	3,500	4,700	5,500	6,050	173
Sheep's milk	1,000 ,,	2,300	3,800	4,000	5,650	246
Eggs	1,000 eggs	5,000	6,000	6,500	6,500	130
Meat	tons	440	505	570	870	193
Total value of produce*	£I.	626,800	870,400	966,750	1,189,350	190
Total value of produce†	£I.	626,800	721,700	885,200	1,902,900	304

* 1949 prices.   † Current prices.

Notwithstanding the increase in livestock production, the value of livestock products has declined relative to gross agricultural production. This must be explained by the official price ceilings imposed upon milk, eggs and meat.

Arab herds of cattle include a considerable number of oxen and

[1] Based upon figures published by the Central Bureau of Statistics and Economic Research, Government of Israel.

steers, which serve as draught animals. Cows, too, are used to draw ploughs. The average annual yield of a cow of local strain does not exceed 500–600 litres. Most milk, cheese and meat consumed by the Arab peasant is produced by sheep.

The production of meat, milk and eggs shows a constant increase from 1948 to 1952.

Poultry-farming as conducted by Jewish farmers is unknown in the Arab sector. According to the Agricultural Census of 1950, there were 74,000 layers on the Arab farms. However, because of economic conditions obtaining in that year—in which much of their produce. was diverted to the Black Market—Arab cultivators rendered lower figures for their poultry.

Arab farmers also engage in bee-keeping, the general trend being in the direction of more modern methods.

In Acre and Jibr-Zerka there are groups of Arab fishermen engaged in coastal fishing.

The Arab village retains traditional features that have become sanctified by many generations. For this reason it has remained a sharply defined sector of the general farm economy. But in view of the economic policy adopted by the Government of Israel after the foundation of the State, and the institution of food-rationing, controls, the maintenance of price levels and the fixing of minimum prices to stimulate agricultural production, it was essential to integrate the Arab economy into the general economic life of the country. Implementation of this policy proved exceedingly difficult. There has always been a considerable disparity between the respective prices commanded by Jewish and Arab produce, because of differences in costs of production. Conditions of free competition, and later the fixing of minimum prices, assured the Arab cultivator of excess profits. This in itself might have merited encouragement had these profits been reinvested and utilized for the development and expansion of the Arab farms, or for raising standards of living in the villages and improving education, sanitation, etc. There were, however, adequate grounds to believe that the standard of living would remain unchanged, and that the surplus profits were being smuggled abroad for various purposes—some of them nefarious, others of questionable value—to the neighbouring countries and were being extracted from the national economy. The Government, therefore, resolved to transfer the marketing of Arab farm produce to a small number of companies upon a basis of differential pricing. This system was based upon computation of costs of production plus fair profit. Different methods were used in calculating costs of Arab and Jewish produce.

The difference between the return obtained for Arab produce at standard prices and the sum the producer was entitled to upon the basis of the above calculation was paid into a special fund of the Treasury, to finance various development works, such as the construction of roads, maintenance of schools, etc.

The system was designed to prevent chaotic market conditions, although in effect it proved impossible to block the leakage of some Arab produce to the Black Market.

It is common knowledge that Arab farmers made very substantial profits during the years 1949–52 when rationing and controls were in force in Israel. Their income from the diversion of part of their produce to the non-competitive Black Market was particularly high. Unfortunately, however, these high returns obtained over a number of years have not left any mark in the Arab villages. New methods of cultivation in a few areas, the introduction of some machinery and fertilizers represent the net result of pressure exerted by the Ministry of Agriculture. Little progress has been made in the Arab sector in the development of agriculture and intensification of cultivation. There is sufficient ground for believing that the money accumulated in the course of these years has been hoarded, partly in this country and partly abroad. In the mountain villages, it must be stressed, where there is little land for field crops, farmers suffer from underemployment. In the absence of work outside the farm the surplus accumulated during the season is used to tide over periods of unemployment.

Arab farming suffered severely during the War of Liberation and it was only after a number of years that 1947 levels of production were achieved. The paucity of the improvements that have been introduced since that year must be attributed, not to a lack of investment funds, but to an inborn conservatism and feelings of political and economic insecurity. The Arab cultivator, while fully conscious of the advantages of producing for the market and endowed with a keen commercial instinct, is extremely cautious in making new investments or even in resolving on new expenditures.

Many years it seems must elapse before the Arab arrives at a full appreciation of the benefits he can derive from modern agriculture, and the investment of his savings in more efficient and profitable means of production.

The State of Israel has opened up new horizons for the development of the Arab farm, for raising standards of living of the Arab peasant and for integrating him into the cultural, political and economic fabric of the country.

5

THE 'MOSHAVA'

THE 'moshava' is a type of agricultural settlement established upon the individual initiative and with the private funds of the settlers. Historically speaking it was the first of the modern Jewish settlements established in this country, and in the nature of things, its methods of cultivation were largely modelled on those of the Arabs, more progressive European techniques being introduced only very gradually. Farming in these 'moshavot' was extensive in the early period and was based upon a relatively large farm-unit necessitating the employment of hired labour. The source of farm labour was the neighbouring Arab village.

With the passage of years each of these 'moshavot' assumed a specific character, shaped by the economic and geographic conditions in which it was placed.

In the Sharon and the Coastal Plain the 'moshavot' were established close to the larger urban centres, the main highways and ports, and near bountiful sources of water. These conditions combined with the competition of extensive Arab farming to induce a process of intensification and even specialization. Almonds, vineyards and, at a later stage, citrus successively became the major or even sole branch of cultivation. Thus a monocultural fruit-growing type of farm developed. The emergence of intensive farming, calling for a substantial investment of capital and much labour, facilitated the absorption of hired labourers both in cultivation and various services. During the Second and Third 'Aliyot' (waves of immigration), in the years preceding the First World War and the period immediately following its conclusion, respectively, most of the newcomers gravitated towards these settlements. The pioneers aspired to live a life of manual labour. The newcomers, however, were a transient element, who worked in these villages only until they were able to settle permanently on the land themselves.

72

Thus two distinct social classes emerged—planters and labourers. There was also a growing class employed in various services, as well as independent small farmers who had no recourse whatever to hired labour.

URBANIZATION OF THE 'MOSHAVA'

As a result of the development of towns, particularly of Tel-Aviv, planters began to engage in urban occupations, without, however, giving up their groves. Gradually they transferred these callings to the 'moshavot' in which they lived, and soon the latter took on a suburban aspect.

The 'moshavot' in the hills of Samaria, Galilee and Judea, unlike their sister-colonies in the plains, have heavy soil, little water and are far from the main arteries of communication. These conditions and a lack of investment capital prevented the intensification of farming. Field crops, cultivated in a manner traditional in the country, improved a little by the introduction of more progressive European methods, together with an expanding area under orchards, constituted the main branches of farming. Here, too, hired labour was common, especially in the vineyards, but at no time did it assume substantial proportions. These villages, accordingly, did not possess the same attraction for new immigrant workers as did the settlements in the plains, while there was little room for the development of an urban or even semi-urban way of life, similar to that which had emerged in the south.

There is no doubt that the principle of private initiative, individual tenure of the soil, employment of hired labour on a considerable scale and the like all contributed to the process of urbanization which overtook the 'moshavot'. However, the major influences operating in this direction were beyond the control of the settlers themselves. They derived from the geographical situation of the 'moshavot', topographical, hydrographical conditions and the like.

An Agricultural Census conducted by the Statistical Department of the Jewish Agency in 1941/42 revealed even then the first signs of a process of urbanization in eight of the larger 'moshavot', Herzlia, Hadera, Ness-Ziona, Pardess-Hannah, Petach-Tikva, Rishon-Lezion, Rehovot and Raanana.[1] Affula and Nathania, which had been counted among the larger 'moshavot', had been planned from the very outset as urban centres, although they, too, possessed certain features of a rural economy.

These eight 'moshavot' occupy extensive areas of land, mostly

[1] D. Gurevich and A. Gertz, *Jewish Agriculture in Palestine*, 1947.

73

under citrus groves, and have a large population, the majority of whom are not engaged in agriculture. They constitute centres for the immediate agricultural district. In Rishon-Lezion and Petach-Tikvah many industrial plants have been established. The process of urbanization has been encouraged by local civic leaders, some of whom were successful, even during the Mandatory period, in obtaining municipal status for their 'moshavot'. Petach-Tikvah, which in 1941/42—the year of the census referred to—had already been recognized as a municipality, continued to be classified as a 'moshava' because of the considerable agricultural property within its boundaries. Shortly afterwards municipal status was awarded to both Rishon-Lezion and Rehovot, after they had, in effect, been towns for some years.

The other 'moshavot' are still agricultural in character. Those in Galilee, under the supervision of the Palestine Jewish Colonization Association, which founded them, continue to engage in the extensive cultivation of field crops, other unirrigated crops and dairy-farming upon rather backward lines. It is only recently that blue-prints have been drafted for the intensification of farming methods and general agricultural development.

The 'moshavot' in Judea and Sharon, in the main, were established by groups of individual settlers, or plantation companies. A few were sponsored by the Palestine Jewish Colonization Association. Until the outbreak of the Second World War, these villages were based upon citriculture. During the war, however, citriculture suffered a major crisis as a result of the impossibility of sending the fruit overseas, and the growers were compelled to seek other means of livelihood. A process of diversification of farming operations was thus set in motion, and the intensive development of dairy-farming, poultry-farming and vegetable-growing, was begun.

POPULATION AND GAINFULLY-OCCUPIED

The census conducted by the Jewish Agency in 1941/42 provides data for the total population, the labour force, the number of farms and the classification of the farms in various categories.

The number of earners who derived their livelihood from agriculture was less than one-third in the larger 'moshavot' (28%) and even in the small 'moshavot' was no more than 42%; 68% of the total population in the 46 'moshavot' and one-half of farms were concentrated in the eight urban 'moshavot'. The classification of these farms, however, varied. In the larger 'moshavot' there were 64% of the citrus orchards and 31% of the intensive mixed farms. The larger 'moshavot', it will be seen, were centres of citriculture.

TABLE 25

Population and manpower in 46 'Moshavot'

No. and type of 'moshava'	Population (souls)	No. of earners	Engaged in Agriculture (souls)	% of earners engaged in Agriculture	No. of privately owned farms	Citrus groves (dunams)	Mixed farms (dunams)
46 'moshavot'	66,823	28,909	9,541	33·0	4,497	2,873	1,526
8* urban 'moshavot'	45,479	19,089	5,342	28·0	2,330	1,840	474
38 small 'moshavot'	21,344	9,820	4,199	42·8	2,167	1,033	1,052

* The eight urban 'moshavot' are: Herzlia, Hadera, Ness-Ziona, Pardess-Hannah, Petach-Tikvah, Rishon-Lezion, Rehovot and Raanana.

During the National Census of 8th November 1948[1] the population of the 'moshavot' was 107,000, indicating an increase of 40,000 in the course of six years. The eight urban 'moshavot' accounted for 25,500 of this increase, Affula and Nathania for 8,642 and the 36 smaller 'moshavot' for 6,400.

The following table gives a comparison of population figures in these ten 'moshavot' for 1941/42 and for 1948.

TABLE 26

Changes in population in 10 'Moshavot'

Year	Herzlia	Hadera	Ness-Ziona	Pardess-Hannah	Petach-Tikvah
1941/42[2]	3,677	6,419	1,496	1,529	14,934
1948[3]	4,693	11,642	2,025	2,569	21,430
Increase	1,016	5,223	529	1,040	6,496

Year	Rishon-Lezion	Rehovot	Raanana	Nathania	Affula
1941/42[2]	6,703	7,512	3,209	3,802	1,650
1948[3]	10,410	12,283	5,922	11,589	2,505
Increase	3,707	4,771	2,713	7,787	855

In the other 'moshavot' the increase in population was as stated— 6,431 (22,323 in 1948; 15,892 in 1941/42).

Upon the basis of an analysis of a 10% sample of the population included in the census of 1948, conducted by the Central Bureau for Statistics and Economic Research, the working population of the

[1] *Statistical Bulletin of Israel*, Vol. I, No. 5, January/February 1950, pp. 388–89
[2] See Note 1, p. 73.
[3] *Government Yearbook 1951/52*, pp. 257–79.

'moshavot' can be classified as follows according to their occupationa structure.[1]

TABLE 27

Occupational structure of the 'Moshavot' (%)

Occupational classes	% Males	% Females	% Total	Economic branches	% Males	% Females	% Total
Primary	23·7	13·1	21·3	Primary	26·6	14·0	23·5
Secondary	36·9	29·9	35·2	Secondary	39·4	31·8	37·6
Tertiary	39·4	57·0	43·5	Tertiary	34·0	54·2	38·9
Total	100·0	100·0	100·0	Total	100·0	100·0	100·0

The same sample provides the following picture of the economic status of the population of the 'moshavot' (in percentages).

TABLE 28

Economic status of the population of the 'Moshavot'

Sex	Workers	Unemployed	Living on capital or rent	Dependants	Total
Males	58·7	0·5	0·3	40·5	100·0
Females	18·3	0·3	0·2	81·2	100·0
Total	38·7	0·4	0·3	60·6	100·0

In 1941/42 33% of the total number of earners derived their livelihood from agriculture. By 1948 this figure had dropped to 21·3%, for of the population growth of 40,000 persons in the 'moshavot' only 16% settled in the 36 rural 'moshavot' and 34,000 persons settled in the ten large, urban 'moshavot', where non-agricultural occupations offered better prospects.

The process of urbanization was intensified in all of these ten urban 'moshavot'. One of them, it will be recalled, Petach-Tikvah, had already been granted township status, while others, including Hadera, Rehovot, Rishon-Lezion and Nathania, had applied for such status at the time and could look forward to obtaining it in the near future.

The sample referred to does not render any data for each individual 'moshava'. However, the Manpower Census of 1950 of the Bureau of Statistics enables us to obtain relevant figures for eight urban 'moshavot'.[2]

[1] Unpublished figures based on the Government Census of population, November 1948.

[2] *Census of Manpower*, 1950, Central Bureau of Statistics and Economic Research, Government of Israel.

TABLE 29

Occupation groups in eight 'Moshavot' (%)

Class of occupation	Herzlia	Hadera	Ness-Ziona	Pardess-Hannah	Petach-Tikvah	Rishon-Lezion	Rehovot	Raanana
Primary	17·5	20·3	26·4	19·8	6·7	7·7	19·8	34·4
Secondary	46·1	34·3	38·9	29·8	49·6	48·2	32·1	30·8
Tertiary	36·4	45·4	34·7	50·4	43·7	44·1	48·1	34·8

In the agricultural occupations Raanana headed the list while Petach-Tikvah came last, followed by Rishon-Lezion, which is rapidly becoming industrialized.

In the tertiary occupations, however, this order is reversed with Petach-Tikvah and Rishon-Lezion at the top of the list.

This table gives a fairly clear picture of the rural or urban character of each of these 'moshavot'. Only in Raanana does agriculture occupy a position of major importance, while in Petach-Tikvah, Rishon-Lezion, Herzlia and Ness-Ziona, the greatest concentration is in the secondary occupations. In Pardess-Hannah, Rehovot and Hadera tertiary occupations are most important. It is probable that the high degree of concentration in the tertiary occupations in Pardess-Hannah—and in Hadera—must be explained by the large immigrant centres situated in their vicinity. In Rehovot, however, the reason is the large number of persons employed in administrative and scientific institutions.

This process of urbanization in these eight urban 'moshavot' is thrown into bolder relief by a comparison with the percentage of the total number of earners engaged in agriculture in 1941/42 and 1950, respectively.

TABLE 30

Earners in agriculture in eight 'Moshavot' (%)

Year	Herzlia	Hadera	Ness-Ziona	Pardess-Hannah	Petach-Tikvah	Rishon-Lezion	Rehovot	Raanana
1941/42	36·6	29·5	29·9	53·4	23·7	22·3	24·2	41·4
1950	17·5	20·3	26·4	19·8	6·7	7·7	19·8	34·4
Absolute change	−19·1	−9·2	−3·5	−33·6	−17·0	−14·6	−4·4	−7·0
Relative change	−52·2	−31·2	−11·7	−62·9	−71·7	−65·5	−18·2	−16·9

In all, it will be seen from the above, the importance of agriculture in the local economy of these 'moshavot' has diminished. It is significant that the greatest relative decline was registered in the 'moshavot' in which the role of agriculture was already on the wane in 1941/42.

The conclusions that may be drawn from the above table corroborate those drawn from the one preceding it, regarding the occupational structure of the population in the eight large 'moshavot'. The population of Petach-Tikvah, the most urbanized, underwent a radical change in respect of the occupations engaged in, compared to other 'moshavot'.

The process of urbanization, which was already marked in the early forties, steadily gathered momentum, particularly after the establishment of the State of Israel and the immense influx of new immigrants which followed.

At the close of 1950 the 'moshavot' had a population of 194,000[1] apparently including the residents of 'maabarot' (Transit Work Camps) in their neighbourhood. The population had thus grown by 87,000 persons since the Census of 1948. The eight large 'moshavot' had 113,000 inhabitants at the end of 1950, or 42,000 more than during the National Census. Other 'moshavot'—excluding Nathania and Affula—increased their population by 30,000.

The following table shows the growth of the large 'moshavot' in the first two years following the establishment of the State of Israel.[1]

TABLE 31

Increase of population in 10 'Moshavot' in 1948–52

Year	Herzlia	Hadera	Ness-Ziona	Pardess-Hannah	Petach-Tikvah
1948	4,693	11,642	2,025	2,569	21,430
1950	12,200	15,000	5,750	4,495	31,500
Growth	7,507	3,358	3,725	1,926	10,070
In %	160·0	28·8	184·0	75·0	47·0

Year	Rishon-Lezion	Rehovot	Raanana	Nathania	Affula
1948	10,410	12,283	5,922	11,589	2,505
1950	18,000	18,200	7,800	23,000	6,202
Growth	7,590	5,917	1,878	11,411	3,697
In %	72·9	48·2	31·7	98·5	147·6

In the other 'moshavot' the population increased from 22,323 to 52,070, or by 29,747 persons.

The rapid growth of Nathania, which in the course of two years doubled its population, is striking. It is of interest to note that while Nathania was established as a 'moshava', the process of urbanization set in at a very early phase in its development, and its agricultural property, at no time very substantial, is no longer of any significance.

[1] Based upon estimates of the Central Bureau of Statistics published in the Israel *Government Yearbook 1951/52*.

Urbanization became more intensified during the Second World War when a number of industrial plants were established within the municipal area. Another factor operating in this same direction has been the development of the town as a vacation and bathing resort. Between 1941 and 1950 the population multiplied sixfold.

Other 'moshavot', the populations of which have increased rapidly, are Petach-Tikvah, Rishon-Lezion, Herzlia and Rehovot.

Since the Manpower Census of 1950 no similar census, capable of reflecting the occupational structure of the 'moshavot', has been carried out. We must accordingly perforce rest content with data for the general growth of population. From the close of 1950 to the close of 1952 the population of the 'moshavot' grew from 194,000 to 201,000 persons. At the close of 1952 the number of residents of the 'maabarot' in the neighbourhood of the 'moshavot' was 48,000. Thus, in effect, the 'moshavot' held a population of some quarter of a million. Meantime plans in regard to the 'maabarot' have changed and they are to be maintained until permanent housing is constructed for the residents in the immediate vicinity. The residents of the 'maabarot' were included in the population figures for 1950, and we shall therefore include them for purposes of comparison with the year 1952. In these two years the population of the 'moshavot' grew by 55,000 persons, of whom the eight urban 'moshavot' account for 35,155, and Nathania and Affula for 6,900. Thus of the total increase of new residents in the 'moshavot', no less than 76·4% were absorbed in the ten large 'moshavot' and only 23·6% in the rural 'moshavot'.

The following table reflects the growth of the 'moshavot'—in the course of 1950–52.[1]

TABLE 32

Increase in population in 10 'Moshavot' in 1950–52

Year	Herzlia	Hadera	Ness-Ziona	Pardess-Hannah	Petach-Tikvah
1950	12,200	15,000	5,750	4,495	31,500
1952	18,600	21,000	9,500	5,500	41,000
Growth	6,400	6,000	3,750	1,005	9,500

Year	Rishon-Lezion	Rehovot	Raanana	Nathania	Affula
1950	18,000	18,200	7,800	23,000	6,202
1952	20,500	23,000	9,000	26,500	9,600
Growth	2,500	4,800	1,200	3,500	3,398

[1] *Government Yearbook, 1951/52* and *1952/53.*

In the other 'moshavot' the population grew by 12,979—from 52,070 to 65,049.

The rapid pace of development of Petach-Tikvah, Herzlia, Hadera and Rehovot is reflected in the above table.

The changes in the population of the ten large 'moshavot' over the decade 1942–52 are striking.

TABLE 33

Changes in population in 10 'Moshavot' in 1942–52

Year	Herzlia	Hadera	Ness-Ziona	Pardess-Hannah	Petach-Tikvah
1942[1]	3,677	6,419	1,496	1,529	14,934
1952[2]	18,600	21,000	9,500	5,500	41,000
Growth	14,923	14,581	8,004	3,971	26,066
Index*	506	327	635	360	275

Year	Rishon-Lezion	Rehovot	Raanana	Nathania	Affula
1942[1]	6,703	7,512	3,209	3,802	1,650
1952[2]	20,500	23,000	9,000	26,500	9,600
Growth	13,797	15,488	5,791	22,698	7,950
Index*	306	306	280	697	582

*1942 = 1000

(In other 'moshavot' the growth in population was 49,157—from 15,892 to 65,049—and the index of population growth was 409 in 1952.)

The above table shows that the largest absolute increase in population was registered by Petach-Tikvah, followed by Nathania, Rehovot, Herzlia, Hadera, Rishon-Lezion, Ness-Ziona, Affula, Raanana and Pardess-Hannah. The greatest relative increase was that of Nathania, followed by Ness-Ziona, Affula, Herzlia, Pardess-Hannah, Hadera, Rishon-Lezion, Rehovot, Raanana and Petach-Tikvah. Of the population increase of 182,500 persons in the period under review, 133,270, or 73%, resided in the ten urban 'moshavot' and 49,150 (27%) in the rural 'moshavot'. It must be borne in mind that these figures included the residents of the 'maabarot' close to the 'moshavot'.

The residents of the 'maabarot' near the ten large 'moshavot', who at the close of 1952 numbered 32,000, were mainly employed in urban occupations and only to a lesser extent in agriculture, as for instance in the picking of citrus. The problem of integrating the

[1] D. Gurevich and A. Gertz, *Jewish Agriculture in Palestine*, 1947.

[2] *Government Yearbook 1952/53*.

residents of the 'maabarot' into the national economy constitutes a subject for a separate study.

AGRICULTURE IN THE 'MOSHAVOT'

In 1942 the 46 'moshavot' occupied a total of 307,000 dunams,[1] of which 245,000 dunams were under cultivation (93,000 dunams irrigated).

In the eight larger 'moshavot' 64,000 dunams out of the total of 84,000 dunams under cultivation were planted with orchards and groves. These eight 'moshavot' cultivated 68% of all the irrigated groves and orchards in the 'moshavot'.

TABLE 34

Cultivated area in 46 'Moshavot' in 1942

Number and type of 'moshava'	Total area (dunams)	Cultivated area (dunams)	Unirrigated orchards (dunams)	Irrigated orchards (dunams)	Orchards as % of cultivated area	Unirrigated orchards as % of total orchards	Irrigated orchards as % of total orchards
46 'moshavot'	307,508	244,815	15,852	92,803	44·4	14·6	85·4
8 large 'moshavot'	102,073	84,442	1,734	62,713	76·3	2·7	97·3
38 small 'moshavot'	205,435	160,373	14,128	60,090	27·6	31·9	68·1

Note: In fact, there were unirrigated orchards in only 27 out of the 46 'moshavot'—mainly in Samaria.

The table indicates the disparity between the large and small 'moshavot' in respect of the crops raised.

Concentration upon citriculture, and the scanty area under other crops, indicates the monocultural character of agriculture in the larger 'moshavot', which indeed was a contributory factor in the process of urbanization. For the purposes of the present study we conducted a survey in 1954 of all of the larger and most of the smaller 'moshavot'. We found that the area occupied by the eight large 'moshavot' had been extended by 81,000 dunams over the past decade; the area under cultivation, however, had not grown.

There are at least 200,000 dunams in the 'moshavot' which are not being cultivated. Even taking into account the sand-dunes of Rishon-Lezion and Hadera, some 150,000 dunams remain uncultivated. This area has been included in the 'Development Areas' set aside for industrial plants, although, to date, progress has not been too considerable.

It must be remembered that the area of land suitable for agriculture in this country is very limited, and importance attaches to every

[1] See Note 1, p. 80.

thousand dunams, especially in the districts in which the larger 'mos-havot' are situated. It would seem far more practical to plant at least 50,000–60,000 dunams of these lands with citrus, as a means of alle-viating unemployment today and of increasing Israel's foreign cur-rency income through exports in the future, than to wait for uncertain industrial development.

LINES OF FUTURE DEVELOPMENT

For the purposes of the survey referred to, local municipal leaders were interviewed. Their views indicate a dual trend: further indus-trial expansion, an accelerated process of urbanization in the 'mos-havot' close to Tel-Aviv; in the more outlying 'moshavot', where the agricultural basis is still strong, it was suggested that there should be development of orchards and perhaps other branches of farming, together with ambitious plans for industrialization. Leaders of the agricultural and semi-agricultural 'moshavot' appear to suffer from an inferiority complex and cannot reconcile themselves to the feeling of lagging behind the larger 'moshavot'.

Most of these local leaders have complained that following the uprooting of the groves in their vicinity the main source of employ-ment had disappeared and they had been compelled to seek substi-tutes. This problem has become more acute as a result of the rapid increase in population. In almost all 'moshavot' large tracts of land have been earmarked for industrial development and adorned with large hoardings reading 'Industrial Development Zone'.

Big industry has not yet succumbed to the blandishments of these signboards. In one important 'moshava' which has been granted an area bordering on the main highway for industrial development, eighteen factories and workshops have been established in the course of the past four years. These, together with the plants which preceded them, employ a total of 320 workers. The largest of these enterprises was established in 1938. But the entire zone does not include a single plant worthy of being called a factory, that is to say, employing at least fifty to sixty workers.

There is one consolation in this process—that most local civic leaders have launched their development projects upon the basis of preliminary planning. This, however, has not prevented the reserva-tion of extensive agricultural areas—which are now desolate—for industries which fail to come.

In one 'moshava' close to the city, occupying 22,000 dunams, in the heart of the citrus belt, only 800 dunams are under groves, while 7,000 dunams are being cultivated temporarily. 14,000 dunams wait for factories. For the purposes of comparison we may recall that when Tel-Aviv numbered 200,000 inhabitants, and possessed a considerable industrial and commercial complex, it occupied no more than 11,000 dunams. A leader of the rural hinterland of the largest of these 'moshava'-towns has complained that the decline in the area under cultivation, despite the grant of supplementary areas, is the direct result of the extension of the municipal boundaries and the creation of new industrial zones.

The uprooting of groves, the parcellation of lands and the sale of plots for housing and other purposes are now common in practically all 'moshavot'. Eloquent testimony to the short-sightedness of growers and landowners, who have destroyed assets worth millions, is provided by the fact that at the same time new lands are being sought for the planting of groves—perhaps even by former citrus growers.

There is also evidence of a lack of proper guidance on the part of the Government. In one 'moshava' complaints have been made against Government opposition to plans for industrial development; the authorities, it appears, favour the establishment of rest-homes, pensions, hotels and vacation centres. It is instructive that one section sought to establish factories, while the other advocated hotels—in an area in which thousands of dunams of groves can be planted providing employment for large numbers of workers and bringing in substantial sums in foreign currency. In these same areas, let it be noted, bananas, sub-tropical and other fruits, and vegetables can flourish. But instead there are a number of hotels, a film studio and some villas which could, with undoubted aesthetic and economic advantage, have been constructed elsewhere.

We found wiser and more intelligent planning in one 'moshava' which seeks to increase the numbers of farmers in its midst, to extend the area planted with groves, because citrus has been and remains the economic foundation of the 'moshava'. Here an industrial area is being planned on the *neighbouring sand-dunes*.

Regional influences, however, are also at work. Thus one long-established 'moshava', which has zealously safeguarded its agricultural character, and has even expanded the various branches of farming in which its inhabitants engage, which was among the last to uproot groves and the first to rehabilitate groves that had suffered from neglect, has also, influenced by the prevailing mood, embarked upon the development of workshops and industries.

The attitude of these local leaders is an expression of their feeling that they are being encouraged to take advantage of prospects of development. They point to the new areas placed at their disposal, the flow of newcomers diverted in their direction, the funds allocated by the Government, the Jewish Agency and other bodies, the housing estates that have been constructed, the roads paved, the parks that have been laid out, the energy and the initiative evinced by the local government authorities and their leaders.

The impression gained, however, is that if all these measures, estimable as they are in themselves, had been preceded by more foresight and wiser planning, if, instead of this blind pursuit of urbanization, the existing healthy agricultural basis had been extended and consolidated, there would be less ground for complaint that the number of breadwinners has not kept pace with the growth of population, that the number of owners of 'kiosks' and small traders who require public assistance is so large, that unemployment is chronic and the expansion of industry so tardy.

If only these same efforts had been invested in rural and not urban development, the larger 'moshavot' would today be capable of making an invaluable contribution towards the restoration of the national economy.

SUMMARY

(1) Three types of 'moshavot' have developed:
- (a) 'Moshavot' which are overwhelmingly urban in character in which only stray vestiges of agriculture have remained—mainly citrus groves which have not yet been uprooted.
- (b) Rural 'moshavot', further from the urban centres, which are based upon farming.
- (c) 'Moshavot' in a stage of transition. In these 'moshavot', agriculture is still an important element in the local economy, though the trend towards urbanization is being encouraged.

(2) Eight large 'moshavot', with an aggregate population of 170,000 inhabitants, occupy an area of 180,000 dunams, of which, however, less than half is under cultivation. The remaining 38 'moshavot', with a population of 100,000, cultivate two-thirds of the 250,000 dunams they occupy.

(3) In none of the three classes of 'moshavot' enumerated are the agricultural potentialities fully exploited.

(4) Land allocated for non-agricultural purposes, such as residential areas, industrial zones, services, etc., appears in all cases to be excessive.

(5) In the urban 'moshavot' and those that are still in the transitory stage, the economic basis is uncertain and the number of persons unable to support themselves by work is inordinately large. By contrast the basis of the rural 'moshavot' is far sounder.

(6) Re-examination of the planning of the 'moshavot' by the national planning authorities is essential. It is also necessary to con-

duct an inquiry into the position of lands within the municipal boundaries of the 'moshavot' and the purposes for which these lands have been earmarked.

(7) The town and village planning authorities must demarcate zones for residential, industrial and agricultural purposes, in each of the 'moshavot', with the object of reducing the area of land allocated for non-agricultural purposes to a minimum.

(8) In view of the restricted area of agricultural land available and the wastage of good land for purposes for which inferior land would do equally well it is essential to prohibit the use of the former other than for agriculture.

(9) All owners of land defined by the competent authority of the Ministry of Agriculture as suitable for citrus-growing must be required to utilize their land for this purpose within a specified term. Should they fail to do so the land must be expropriated in favour of the Minister of Agriculture, who will lease the area to any person or corporation or to the National Orchards Company, ready to undertake the development of citrus groves. The object of this measure must be to build up a reserve of land, with the object of restoring citrus-growing to the place it formerly occupied in the national economy.

Far-reaching and rigorous measures of this nature are justified in view of the comparatively restricted zone suitable for the cultivation of citrus.

(10) With a view to preventing over-concentration in urban centres and to securing the widest possible distribution of population throughout the country, it is highly desirable that all classes of 'moshavot' be planned upon a basis of suitable integration of industry and agriculture. The national financial and planning authorities must use their powers to achieve these objectives.

(11) The 'moshava' as a type of settlement can and must continue to fulfil the function for which it was originally planned, provided its agricultural basis be strengthened and extended.

6

MIDDLE-CLASS SETTLEMENTS

ACCORDING to the classification conventional in Israel, farm-owners in the 'moshavot' are included in the category of 'private settlers', as distinct from 'kibbutzim' (collectives) and the 'moshavim' (co-operative villages) which together comprise the 'workers' settlements'. However, there exists another, intermediary category of village organization, the members of which, while affiliated neither to the private nor the workers' settlements, nevertheless possess characteristics associated with both.

The wave of immigration into Palestine in the early twenties of the present century, emanating mainly from Poland, brought a considerable number of settlers belonging to the so-called middle-class. Some of these immigrants decided to engage in agriculture, concentrating mainly on citrus-growing. They provided the initial stimulus for the development of the 'moshavot' in the Sharon Valley. In Hefer Valley, too, a number of villages were founded by these middle-class settlers. However, it was only rather more than a decade later, when settlers possessed of some capital of their own began to arrive from Germany, Austria and Czecho-Slovakia in larger numbers, that their specific contribution towards village organization in Palestine, the 'kfar shitufi' (co-operative village), took tangible shape. About one thousand of these immigrants settled on the land, one-third of them in existing villages, the remainder establishing settlements of their own.

This phase of agricultural settlement differed from all those which had preceded it. Mostly these Central European immigrants were older than the normal run of settler. Sociologically, they came from the more prosperous classes, and comprised a relatively high proportion of former members of the liberal professions and other persons of academic education. Another important characteristic distinguishing them from the 'chalutzim' (pioneers)—they came accompanied by their families. They brought with them some capital, which though not large in comparison to their previous wealth, was nevertheless adequate for the development of a smallholding. Very few of them had any previous experience or training in agriculture. In spite of this, with the exception of small loans, extended by the Jewish national institutions, they relied entirely upon their own efforts and resources.

Ramot Hashavim was the first of the villages founded by these settlers, followed in close succession by Kfar Bialik, Kfar Yedidia, Gan Hashomron and Naharia. In all, twelve settlements of this type were established. In 1935 the Rural and Suburban Settlement Corporation (usually referred to as RASSCO) was floated with the express purpose of aiding these settlers. RASSCO was responsible for the establishment of Kfar Shmaryahu, Sdeh Warburg, Beth Yitzchak and Shavei Zion. The Jewish Agency for Palestine created a special Division of its Settlement Department, the function of which was to foster middle-class settlement on the land.

The main feature distinguishing these villages from the 'moshavot' is that they are based upon mixed farming and the labour of the farmer and his family. Indeed theirs is a typical family farm not differing, essentially, from family farms in other countries. But unlike the 'moshav ovdim' self-labour is not one of its basic principles and both in theory and in fact hired labour is employed—as in England or in America. As the name—'kfar shitufi'—indicates, co-operation is basic in the organization of these villages, and in each there is an Agricultural Co-operative Society, which controls the purchase of equipment and supplies, and the sale of farm produce, besides undertaking the maintenance of diverse public services in the village. Indubitably to a very considerable extent the success of these settlers, despite their comparatively advanced age and lack of any agricultural experience, must be attributed to their excellent compact co-operative organization.

Another difference between these middle-class settlements and the 'moshvei ovdim' is the size of the farm-holding. In all villages established on privately-owned land the holdings are tiny, and for that reason indeed the settlers engage principally in the breeding of poultry for egg-production. This monocultural contribution has produced high standards of efficiency, and Ramot Hashavim, the first of these settlements, has developed into a model poultry-breeding village.

Shavei Zion, another village founded by settlers of this class, is out of the common run. The original group of settlers comprised some dozens of families who while still in Germany had engaged in farming. They resolved to transfer their community, and whatever assets and property they could salvage, en bloc to Palestine, organizing upon their own initiative a 'moshav shitufi' without indeed knowing anything at all of the existence of similar villages in this country. Their motives in this step were entirely practical. The assets of the individual members were placed in a common fund. Their village was based upon community of property and collective production, housing and consumption being organized on individualistic lines. The village succeeded in striking root, and today is one of the most efficiently run in the country. After the establishment of the State of Israel, Shavei Zion absorbed a number of new settlers.

Naharia represents another interesting type of middle-class settlement. The first group of settlers aspired to create a village of smallholdings no more than a few dunams in extent. In the course of time, however, Naharia has developed into a thriving vacation resort, exploiting to the full its situation on the coast and other natural advantages. The farmers of the community are organized in the 'Smallholders Agricultural Co-operative Society of Naharia, Ltd.' Naharia, too, experienced a period of expansion following the establishment of the State of Israel.

Most farm-holdings in the middle-class settlements are small, ranging from seven to seventeen dunams. The limited extent of the farm has provided the impetus for intensive cultivation, though the smallness of the villages themselves has retarded progress in other directions. The establishment of the State enabled both an expansion of the land-holding to an average of twenty-five dunams, and an increase in the number of settlers in each village.

Beth Yitzchak provides another instructive example of a thriving middle-class village. The farms of the original sixty settlers, at first no more than seven dunams in extent, have in the course of the years been doubled in size. A new phase in the development of the village opened up with its merger with the two neighbouring settlements of Nira and Shaar Hefer, and the absorption of fifty new settlers. At the time of writing (1955) the village co-operative society controls 4,800 dunams of land, of which 3,000 are under irrigation. The livestock of the 185 farms included 445 head of cattle and 70,000 chickens. One thousand dunams were under vegetables. The annual value of farm produce was of a magnitude of £I.1·5 million.

In this context mention must also be made of the hundreds of small-farmers in the 'moshavot' who have been organized in the Agra Co-operative Societies, under the aegis of the Jewish Agency's Division for Middle-Class Settlement.

The function of these co-operative societies is to organize the marketing of farm produce, as well as to purchase agricultural equipment and supplies. The Agra societies have combined with the co-operative villages to form a national body known as 'The Agricultural Council' which today represents most of the middle-class settlements and also engages in the organization of new groups of settlers.

An organic phase in this process of development, following the consolidation of the villages and the growth of co-operative institutions, has been the creation of a national co-operative institution. The 'Tenneh' Co-operative Society, founded in 1940 by the village co-operatives to centralize and co-ordinate the sale of their produce, was modelled upon the lines of the 'Tnuvah' Co-operative Society which serves the workers' settlements. 'Tenneh', to which forty-eight village societies are affiliated, operates a central creamery in Kfar Shmaryahu.

The 'Haspaka' Corporation is the central purchasing agency of the middle-class villages and fifty-one affiliated societies. In the 1954/55 financial year 'Haspaka's' turnover topped £I.6·5 million. These villages and co-operatives as well as the national bodies they have created are organized in the Bachan Audit Union, which advises them on various administrative and organizational matters and audits their books. One hundred and four co-operatives and other bodies are affiliated to Bachan.

The middle-class co-operative settlements constitute a distinct segment of Israel's agricultural economy. In 1947 they were responsible for about 20% of the country's agricultural production, their contribution being particularly high in egg and vegetable raising.

During the mass influx of immigration which set in after the establishment of the State of Israel, the middle-class settlers were of a different character. They came with little capital of their own, and officially all immigrants who could raise one-half of the investment required for settlement on the land were classified as middle-class settlers. In practice, however, it was found that as a result of the prevailing economic instability, and especially because of rising prices, the funds at the disposal of most of these settlers constituted hardly more than one-third of the investment capital required. The balance was made up by an allocation from the Jewish Agency, the Israel Government providing the necessary housing. The farm-unit has in the meantime been increased to 25–28 dunams of irrigated land.

In recent years middle-class settlement has taken a less favourable turn. The dimensions of poultry-breeding, their main branch of farming, in which they excelled, have been restricted in conformity with the Government's agricultural policy. Concentration upon market-gardening, to which they turned as an alternative, led to over-production and concomitantly to a sharp decline in prices. In addition, the new settlers in these villages proved of a more heterogeneous character and less amenable to discipline. In spite of these difficulties,

however, the complex of middle-class settlements has registered gratifying expansion. In the six years after the establishment of the State,
1,200 newcomers were settled in such villages, roughly about half of
them founding new settlements. The new settlers invested some £1.3
million of their own capital. Each settler in this category was provided
with a dwelling, having a total floorspace of 50–60 square metres,
and comprising a hall, two rooms and the usual services and conveniences. The area covered by the irrigation network is 15,000
dunams in extent. Dairying has been expanded, while the raising of
chickens for slaughter has given a new stimulus to poultry-breeding.
The expansion of arable-farming and the establishment of tractor
stations has provided valuable aid for those poultry farmers wanting
the land to raise their own feed. Citriculture, too, is now an important
branch in these farms. Eighteen hundred dunams of fruit-bearing
groves have been made over to middle-class settlements, and 2,500
dunams have been planted.

The absorption of new settlers in the existing villages has facilitated the integration of the former in their new way of life, and has
accelerated the pace of agricultural development. Furthermore the
enlargement of villages has redounded to the advantage of both
new and old settlers. Givat Ada, a small village of thirty farms in
the Samaria district, which after stagnating for decades, experienced
a new era of prosperity following the absorption of ninety new
settlers, provides a typical example of this process.

The fourteen new villages founded in this period are scattered
throughout Israel, from Western Galilee to the Coastal Plain. The
settlers come from many parts of the world—England, Egypt, Iraq,
Poland, Argentine, being among the countries represented. Like most
other settlements in Israel they constitute a microcosm of the ingathering of the exiles that is a feature of the state, with a strong admixture of veteran citizens of the country, and native-born Israelis,
though new immigrants account for at least 70% of the total
population.

Economic conditions in these villages are not uniform. Some have
already succeeded in striking root and are well on the way to prosperity. Others, however, still lack essential equipment and other
means of production. Of the 1,200 settlers in this category who were
placed on the land, 600 have already entered upon the 'consolidation'
stage, in other words they are receiving the final instalment of their
settlement grant, though they still require a measure of assistance to
achieve full economic independence.

Five new middle-class villages in the Bashit area have been proclaimed a region block, with a District Council exercising jurisdiction
in the economic and cultural spheres.

Experience gained in the promotion of this type of settlement has prompted the Jewish Agency's Middle-Class Settlement Division, in co-operation with RASSCO, to undertake the preliminary development of farm-holdings, so that henceforth the settlers will find ready on arrival, a dwelling, farm-buildings, livestock, ten dunams of improved land with an irrigation network installed and a young citrus grove. In 1955, 600 such farm-units were being developed, 300 of them in existing villages and the remainder in new settlements, from Taanach in the North, to Ascalon in the Southern District.[1]

[1] (a) Reports prepared by Dr. L. Pinner, Director of the Middle-Class Settlement Division of the Jewish Agency for Palestine. (b) Annual reports of the Bachan Audit Union.

7

WORKERS' SETTLEMENTS

A N interesting process of mutual inspiration of social and economic ideas was a characteristic feature in the evolution of the
early 'kibbutzim' and 'moshavim'. There seems little doubt
that the pioneers of the first co-operative villages—the 'moshavim'—
for whom the highly individualistic structure of the 'moshavot' was
incongenial, were strongly influenced by the collectivism of the
'kibbutzim'.

It is widely held among agrarian sociologists that preference for
either individualistic or co-operative forms of social organization is a
function of national character. It is upon this assumption that they
seek to explain diverse forms of land tenure in various countries.
The 'obshtshina' or 'mir' system was common in eastern Russia; in
the western provinces of that country, however, it was unknown,
and probably influenced by Western European ideas—shaped in
turn by the system of land ownership recognized by Roman Law—
land there was privately owned.

The United States Congress, convened in the spring of 1789—
the first to be held under the new constitution—proclaimed an agrarian policy which many writers on American internal policies regard
as without precedent in the history of Mankind.

One of the questions raised by this policy was:

'Will the new nation follow patterns deeply rooted in the cultural
heritage derived from its European forebears, or strike out boldly
into paths peculiarly its own? Will it continue or expand an aristoc-
racy of owners of manorial estates on the English pattern? Will it
become wholly a master-and-slave economy? Or will it turn toward
a semi-egalitarian programme based on strongly rooted property

rights in land, and a rugged, deeply democratic society of shirt-sleeved farmers?"[1]

The American pioneers sought to solve the problems of ownership of the soil in accord with their distinctive ideas. In effect, however, these problems remain unsolved to the present day.

To the Jews of Palestine belongs the signal distinction of being the first—and, in the opinion of many, still the only—nation in the world to prove that the land problem can be solved by national ownership. The principle of national ownership indeed has guided Jewish agriculture in this country along distinctive channels. Moreover, the Palestinian pioneers fortified this principle by consistent application of another, that of self-labour, since the foundation of the first co-operative settlements over forty years ago. Had these twin tenets of national ownership of the soil and self-labour been adopted by other nations, the prospect of Mankind would be happier than it is today. In Israel, certainly, their translation into daily life has produced notable achievements in many spheres. The architects of the 'kibbutz' and the 'moshav' aspired to realize their social, economic and cultural ideals. In the central design and master plan of the edifice they envisaged, they were inspired by a common source. It was in the minor architectural details that they diverged. Some sought to do away with all divisions in social life, to accomplish a maximum degree of co-operation in all fields of production and distribution and the care of the children. Others, regarding the family as the basic social and economic unit, preferred a more individualistic way of life. There has even emerged a third, hybrid form, the 'moshav shitufi', seeking to combine features of the 'kibbutz', such as collective production, with those of the 'moshav', notably individual consumption of the product of labour. We shall study this form too, which has afforded much moral satisfaction to its adherents, within the general framework of the 'Moshav' Movement.

Thus three types of workers' settlements, each distinguished by its special character and way of life, have developed. Today we can see how unjustified were the complaints made at one time by leaders of the 'Moshav' Movement that the 'kibbutzim' were being favoured in accommodating and training young immigrants of the Youth 'Aliyah'. Indeed, any similar charge today—were it to be uttered—of undue preference for the 'moshavim' in the absorption of new immigrants would be equally unwarranted.

In the last resort the choice between maximum collectivism and individualism depends upon the character, the natural inclinations

[1] Murray R. Benedict, *Farm Policies of the United States 1790–1950*, New York, 1953, p. 3.

and the training of the settler. Even as late as the eighteenth century, within the young American nation there were still conflicts between the various immigrant groups, deriving from their countries of origin. The situation in Israel at various periods has been not dissimilar. Newcomers to Israel, coming from different environments sought types of settlement most congenial to their specific outlook, ideals and education. The settlers of the Second and Third 'Aliyot' (waves of immigration) who reached this country, impelled by the national and social urges of the revolutionary periods of 1906 and 1917, differed fundamentally from the immigrants who came many years later after the destruction of the European Jewish communities and the crisis that had overtaken the Jewries of the North African and Middle Eastern countries. Differences of background and motivation of emigration to Israel, have profoundly influenced their way back to the land.

THE FOUNDATION OF THE NEW SETTLEMENTS

In October 1947 there were in Palestine 72 'moshavim' and 145 'kibbutzim' with a population of 18,268 and 43,258 respectively,[1] the ratio being 257 : 100 in favour of the 'kibbutzim'. In October 1953 the population of these settlements[2] had grown to 22,045 in the 'moshavim' and 59,651 in the 'kibbutzim',[3] the ratio now being 271 : 100. The number of residents in the 'moshavim' included in the survey of 1947 had increased by 4,000 and those of the 'kibbutzim' by 16,400. Naturally the disparity in the absorptive capacity of each of these types of settlement must be taken into account, for the 'moshav' is a closed settlement and the 'kibbutz' an open one. Less than half of the population growth in the 'moshavim' is accounted for by the arrival of parents of settlers and an increase in the number of children. The remainder are new settlers who took up vacant holdings, or were allowed to settle in the existing 'moshavim' following a redistribution of lands. But the difference in this respect between the two types of settlement is conspicuously illustrated by the number of new villages established during 1947–53; 24 'moshvei ovdim', 11 'moshavim shitufiim' and 155 'moshvei olim' were founded in this period, bringing the total number of 'moshavim' of all types up to 250, with an aggregate population of 74,588. The parallel

[1] *Statistical Bulletin of the Audit Union for Workers' Agricultural Co-operation Ltd.*

[2] The number of 'moshavim' had decreased by twelve and of 'kibbutzim' by six, partly as a result of evacuation during the War of Liberation, and partly because of reorganization.

[3] The above-mentioned *Statistical Bulletin*, 1953/54.

figures for the 'kibbutzim' were 77 new settlements, making a total of 216 with a population of 71,569. In the course of these five years the 'Moshav' Movement grew by 52,000 persons and the 'Kibbutz' Movement by only 12,000.

These figures reflect a veritable revolution. The number of 'moshavim' has grown in a proportion of 100 : 347. The increase in population has been even more striking—100 : 408.

Figures for 'kibbutzim' show a 49% increase in the number of settlements and 66% in population.[1]

The relative importance of the 'moshav' and the 'kibbutz' is reflected in the percentage of the total population of Israel resident in each type of settlement.

TABLE 35

Relative weight of the 'Moshav' and the 'Kibbutz',
as % of population

Year	'Moshav' population as % of total population of Israel	'Kibbutz' population as % of total population of Israel
1947	3·0	7·1
1952	4·7	4·9

The 'kibbutz' has not yet lost its position of predominance but while the relative importance of the 'moshav' has risen by 57% that of the 'kibbutz' has declined by 31%.

THE 'MOSHAV' AND THE 'KIBBUTZ'—A COMPARISON AND
A CONTRAST

It is not the purpose of the present study to discuss the differences between the 'moshav' and the 'kibbutz'. Together they have borne the brunt of the colonization of the country, throughout the first half of the present century. Both in this respect and in the broader considerations of agrarian policy they have proved faithful and efficacious instruments, complementing each other in the supreme task of settling immigrants on the land and restoring a nation of exiles to tillage of the soil. Neither, in the absence of the other, could have accomplished what was achieved by both.

The sharp controversy, which sporadically has been carried on by the more zealous adherents of the 'kibbutz' and the 'moshav', has now died down. Settlers, planners, sociologists and economists all prefer to discuss each separately and to examine the peculiar characteristics, the structural, economic, social and other problems of each in its own context.

[1] Work 'kibbutzim' and 'irgunim' have not been included.

95

A number of problems common to both the 'moshav' and the 'kibbutz' type of settlement, or specific to either of them, will be analysed here. This method of comparison and contrast, we feel, possesses distinct advantages in clarifying the points at issue.

TABLE 36

Comparative changes in cultivated area[1]

| | Cultivated area (physical) | | | | | Area cultivated by settlements founded prior to 1947 | | |
| | 1946/47 | | 1952/53 | | | 1952/53 | | |
	Dunams	Index	Dunams	Index	Index 1947=100	Dunams	Index	Index 1947=100
'Moshavim'								
Total area	151,710	100	673,420	100	444	315,371	100	208
Under irrigation	38,397	25	132,698	19	346	66,766	21	174
'Kibbutzim'								
Total area	331,520	100	1,385,605	100	418	1,021,383	100	308
Under irrigation	72,625	22	198,271	14	273	165,414	16	228

The area of land cultivated by the 'moshavim' founded prior to 1947 almost doubled in the six years under review, increasing by 160,000 dunams; that cultivated by the 'kibbutzim' grew more than threefold, by 690,000 dunams—over half a million dunams more than in the 'moshavim'.

Expansion of the irrigated area was also greater in these 'kibbutzim' than in contemporary 'moshavim'. In absolute figures the 'moshavim' extended the area under irrigated cultivation by 28,000 dunams as against 93,000 dunams in the 'kibbutzim'.

Despite the marked preference shown by new immigrants for the 'moshav' type of settlement and the increase—by 344%—in the cultivated area, the 'kibbutzim' in general maintained their position and registered an increase of 318%. The 'moshavim' brought additional 520,000 dunams under cultivation in this period; the parallel figure of the 'kibbutzim' was over one million dunams.

In 1953 the 'kibbutzim' cultivated an area double that of the 'moshavim', namely somewhat less than half of the gross cultivated area in the country as compared with rather more than one-fifth of the total cultivated by the 'moshavim'. The relation between the area cultivated by each of the two types of settlement was 206 : 100.

It is too early yet to draw any conclusions in regard to the area under cultivation. It must be borne in mind that in these six years the 'kibbutzim' cultivated lands far distant from their own settlements,

[1] See Notes 1 and 3, p. 94.

on a temporary basis. These lands are being held in reserve for new villages of all types to be established in the future.

UNIT OF LAND

There is no national average which can constitute a standard in our discussion of the unit of land. We shall revert, for this purpose, to the method adopted in a previous chapter dealing with land problems, and analyse a sample of 'kibbutzim' and 'moshavim' situated in various parts of the country. In each district we shall select two groups of typical settlements comprising three or four 'moshavim' or 'kibbutzim' from the same neighbourhood, where climatic and soil conditions are more or less uniform.

TABLE 37

Comparative size of farm-units[1] (dunams)

	1947				1952			
	Per unit		Per capita		Per unit		Per capita	
	P*	T	P	T	P	T	P	T
(A) Gilboa District								
'Moshav'	100·0	—	15·4	—	71·4	41·0	11·3	6·5
'Kibbutz'	66·0	1·4	17·7	0·4	45·7	35·5	13·1	10·2
(B) Western Jezreel								
'Moshav'	110·0	—	14·4	—	115·5	13·0	13·0	1·5
'Kibbutz'	47·3	7·3	11·8	1·8	31·3	33·3	7·9	8·4
(C) Hefer Valley								
'Moshav'	25·7	4·1	4·3	0·7	35·0	40·0	5·9	6·3
'Kibbutz'	11·8	2·2	3·1	0·6	12·6	16·8	3·1	4·2
(D) Southern District								
'Moshav'	50·5	—	8·0	—	45·0	11·8	6·4	16·7
'Kibbutz'	7·2	6·2	1·9	7·6	22·1	28·7	5·3	6·9

* The letters P and T indicate land held permanently and temporarily respectively. The figures represent the average holding of each group in the district.

In calculating the per capita land holding in the 'kibbutzim' we have included the following categories of local residents: members, candidates for membership, local children, parents of settlers and other dependants, members and candidates for membership temporarily resident elsewhere.

The method of calculating the land-holding per unit may be reduced to the following formula: $\dfrac{M + C + Mo + Co}{2}$, where M =

[1] Based upon balance sheets of the settlements used in our sampling and a.m. *Statistical Bulletins of the Audit Union.*

members, C = candidates, Mo and Co = members and candidates temporarily absent from the settlement.

In regard to the 'moshavim', however, a different method has been used. The 'moshav', as already stated, is based upon the family holding. In computing the per capita land unit we have included the entire permanent population of the village, namely, settlers, children and other dependants and professional and skilled workers (including their dependants) resident in the 'moshav' as well as all permanent residents temporarily absent. This method has perforce been adopted as comparison between the two types of settlement on the basis of farm-units is not suitable. In the 'moshav' the number of units is identical with the number of farms, while in the 'kibbutz' every adult couple including those not engaged in agriculture is taken as representing a farm-unit. We believe that the method adopted provides a more satisfactory basis for comparison than would, say, reckoning, in the 'moshav' as in the 'kibbutz', every adult couple as a farm-unit.

This calculation indicates the substantial proportion of non-agricultural residents, including skilled and professional personnel, in the 'moshavim'. The following table reflects the proportion of permanent non-agricultural residents of this category in the selected 'moshavim':

TABLE 38

Non-agricultural residents in 'Moshavim' (%)

District	1947	1952
(A) Gilboa	21	27
(B) Western Jezreel	30	36
(C) Hefer Valley	20	21
(D) Southern District	27	37

These permanent residents of the 'moshavim' who do not engage in farming include various public officials, independent professional men, artisans, etc. The public officials include the secretary of the village, the accountant, workers in the co-operative store and creamery, teachers, etc., who are salaried employees of the Village Committee, their remuneration being based upon the rate officially recognized in their trade or profession. Independent tradesmen, including shoemakers, carpenters, blacksmiths, drivers of privately owned lorries, etc., are paid individually by their customers. In the majority of cases members of both classes of these residents have smallholdings, often of substantial proportions.

There are also residents who because of their employment elsewhere are no longer members of the village, but who for various reasons continue to reside in it, with the consent of the Village Committee. Of course in the 'kibbutz', teachers, shoemakers, carpenters and others engaged in non-agricultural occupations are full and equal members of the settlement.

Superficially it would appear that whatever common basis there is for comparison between the two types of settlement is very narrow. Closer study, however, reveals that within the limits of these reservations there are many features of economic development which they share. (Henceforth we shall refer briefly to the respective districts as A, B, C and D, in the order in which they appear in the foregoing tables.)

Two 'moshavim', situated in Districts B and D respectively, belong to the category of 41 veteran 'moshavim' referred to elsewhere, in which the proportion of residents, who are not members of the village, is highest. The average proportion of such residents in 43 veteran 'moshavim' was 19% in 1947. In 1952 the percentage for 41 'moshavim' (without Atarot and Ben-Shemen) had declined to 18%.

To facilitate comparison we shall present the permanent land-holding per capita in the form of an index.

TABLE 39

Comparison of per farm and per capita land-units
(indices)[1]

District	Settlement	1947		1952	
		Per unit	Per capita	Per unit	Per capita
A	'Moshav'	100	100	100	100
	'Kibbutz'	66	115	64	116
B	'Moshav'	100	100	100	100
	'Kibbutz'	43	82	27	61
C	'Moshav'	100	100	100	100
	'Kibbutz'	46	72	36	53
D	'Moshav'	100	100	100	100
	'Kibbutz'	14	24	49	83

The 'Kibbutz' in the Southern District operates factories and workshops. The fact that subsequent to the War of Liberation the land-holding of this settlement was increased threefold is an indication of the acute shortage of land from which it suffered in previous years, and which indeed stimulated the development of non-agricultural enterprises. But even if the number of members employed in these branches is taken into account and the aggregate of farm-units

[1] Based upon balance sheets of the settlements used in our sampling and a.m. *Statistical Bulletins of the Audit Union.*

reduced accordingly, the land holding remains far smaller than that of the 'moshav' in the same vicinity.

The assumption that the land-unit per family, per worker and per capita of population is smaller in the 'kibbutz' than in the 'moshav' is borne out by a comparison between the 'moshav' and the 'kibbutz' situated in the Western Jezreel District. The latter—in 1947—had no industrial enterprise. The 'kibbutz' in the Gilboa District, too, does not operate any industries, but a number of its members are employed in the regional creameries, etc., of the 'Tnuva' Corporation, in the district. The reduction of the number of farm-units as a result would not materially alter the picture.

Increasing use of irrigation in the 'kibbutzim' has had the effect on still further decreasing the per capita land-unit.

The examples cited above are all of older-established settlements. We shall now analyse the situation in younger settlements, four 'moshavim'—of which two are immigrants settlements—and four 'kibbutzim' all of about the same age, and in the same districts as the older settlements already dealt with.

The proportion of the non-agricultural workers in the four young 'moshavim'—in the order of the districts enumerated—was 16%, 10%, 2% and 12% respectively. In the following table, too, the number of skilled tradesmen is taken into account, in calculating the per capita land-unit.

TABLE 40

Comparative land-units in young settlements[1]

District	Type of settlement	Dunams per unit		Dunams per capita		Index	
		P*	T*	P	T	Per unit	Per capita
A	'Moshav'	88·1	—	22·0	—	100	100
	'Kibbutz'	79·8	—	32·4	—	91	147
B	'Moshav'	39·8	45·4	11·1	12·7	100	100
	'Kibbutz'	—	59·1	—	18·6	69	78
C	'Moshav'	—	33·7	—	9·7	100	100
	'Kibbutz'	19·6	73·4	7·2	26·7	58	79
D	'Moshav'	—	23·9	—	5·9	100	100
	'Kibbutz'	—	53·6	—	14·9	224	254

* P and T represent land held permanently and temporarily respectively.

No significance attaches to the fact that the land has been registered as temporarily held. This is due solely to a formal delay in the transfer of the land to the ownership of the Jewish National Fund. For the purposes of this study we have regarded these lands as being

[1] Based upon balance sheets of the settlements used in our sampling and a.m. *Statistical Bulletins of the Audit Union*.

cultivated on a permanent basis by the settlements. With the sole exception of the 'kibbutz' in District C these lands, or at any rate the major part of them, will be transferred to the settlements. The 'kibbutz' in District C cultivates land at some distance. Only in this case will the land be regarded as held temporarily. The 'kibbutz' in District D also cultivates outlying lands, but seeing that so far no land has been transferred to it for permanent cultivation we shall not distinguish between land permanently and temporarily cultivated.

The large area of land allocated to the young 'kibbutzim', in comparison with the older-established 'kibbutzim', is explained by the fact that the former have not yet got their full complement of settlers, despite the fact that for the purposes of our sample we have chosen 'kibbutzim' with larger populations.

In the relatively short period since their foundation these younger settlements have not succeeded in consolidating their farms to an extent enabling comparison, though certain economic processes can already be distinguished. The progress made in development is already clear.

It must also be noted that the 'moshav' in district A had no land whatever under irrigation, while the 'kibbutz' in the same district had only 60 dunams. Of the settlements in District B the 'moshav' had 10·5 dunams under irrigation per unit and the 'kibbutz' 12·3 dunams. The 'kibbutz' was accordingly more favourably placed in this respect, but taking into account the land temporarily cultivated by the 'moshav' and the advantage possessed by the 'kibbutz' in respect of the irrigated area, the per capita land-unit of the latter is considerably larger.

In District C the 'moshav' had 6·3 irrigated dunams per unit as against the 'kibbutz's' 4 dunams. But here the area of land held temporarily by the 'kibbutz' is far larger.

In District D the 'moshav' had 5·5 dunams under irrigation per unit and the 'kibbutz' 5·1 dunams. In regard to land under irrigation per capita of population, the similarity is even more marked—1·5 dunams in the 'moshav' as against 1·4 dunams in the 'kibbutz'. In respect of the land-holding the 'kibbutz' is in a more favourable situation, having more than half as much land again as the neighbouring 'moshav'.

In addition to the relation between land and population it is instructive to examine comparative data indicating other aspects of agricultural development, such as the unit of irrigated land, irrigated fodder crops, head of cattle, and laying birds and mechanization (as represented by the number of tractors) per capita. Besides rendering the number of tractors per capita, we shall compute the area cultivated per tractor.

For our purpose we shall once again select four 'moshavim' and four 'kibbutzim' in the same districts (A—Gilboa; B—Western Jezreel; C—Hefer Valley; D—Southern District).

TABLE 41

Comparison, of some per capita units in older 'Moshav'[1] and 'Kibbutz' in different districts

District	Settlement	Irrigated orchard per capita (dunams)	Irrigated fodder per capita (dunams)	No. of cattle per capita	No. of laying hens per capita	Tractors per capita	No. of dunams cultivated per tractor
A	'Moshav'	0·19	1·77	1·46	43	0·021	845
	'Kibbutz'	0·68	1·26	0·65	19	0·031	727
B	'Moshav'	0·51	1·75	1·50	34	0·044	309
	'Kibbutz'	0·70	0·44	0·52	16	0·024	616
C	'Moshav'	0·83	2·72	1·69	34	0·011	1156
	'Kibbutz'	0·29	0·61	0·27	1·2	0·016	370
D	'Moshav'	0·25	2·43	1·66	35	0·029	783
	'Kibbutz'	0·42	0·49	0·22	4	0·015	830

The above table indicates:

(1) The quota of irrigated orchards in the older settlements ranges between a fifth and four-fifths of a dunam per capita. In one settlement orchards cover 488 dunams, in another—574, and in a third as much as 750 dunams. The importance of fruit-growing in the agricultural economy of the older settlements is obvious. In this respect the difference between the 'moshavim' and the 'kibbutzim' is not considerable, with the sole exception of Hefer Valley where the unit of orchards in the 'moshav' exceeds that in the 'kibbutz' threefold.

In the Southern District and the Western Jezreel Valley the quotas are practically equal.

(2) The area under irrigated fodder crops shows the dimensions of dairying. In all areas irrigated fodder crops are cultivated to a greater extent by the 'moshavim' than by the 'kibbutzim'.

(3) Upon a per capita basis dairying in the 'moshav' in the Gilboa District is twice as important in the village economy as in the 'kibbutz' in the same district, whose dairy herd has enjoyed a considerable reputation for several decades. In Western Jezreel too, the number of cattle per capita in the 'moshav' is three times that in the neighbouring 'kvutza', though the latter is one of the most highly developed collective settlements in Israel. In Hefer Valley and the Southern District the dairies of the 'moshavim' are six and seven times as large as those of the 'kibbutzim'.

[1] Based upon balance sheets of the settlements used in our sampling and a.m. *Statistical Bulletins of the Audit Union.*

(4) Poultry farming. Poultry rearing is almost uniformly developed in the 'moshavim', the optimum size of the run being 200–250 birds. In the 'kibbutzim', however, other considerations operate, such as the quantity of domestically grown feeding grains, suitable personnel and availability of investment capital (which is now an important factor, following the introduction of the American type of run, housing 2,000–3,000 birds). Our table indicates that even in the 'kibbutz' in Hefer Valley, which maintains a poultry run solely to meet domestic consumption, flocks comprise 1,360 hens and 3,600 pullets. Neither does the 'kibbutz' in the Southern District market large quantities of eggs. In general 'kibbutzim' in Israel rarely average more than fifty to sixty laying hens per farm-unit. Poultry farming, like dairying, is more highly developed in the 'moshav' than in the 'kibbutz'.

(5) Mechanization. Since the introduction of agricultural machinery into Israel, it has been widely held that mechanization has been carried to excessive lengths in the 'kibbutz', contrary to the 'moshavim', where draught animals are still preferred. In the four moshavim included in the accompanying table the number of draught animals was 64, 114, 151 and 93, respectively, as compared with 7, 8, 3 and 25 animals in the 'kibbutzim' in the same districts. In regard to tractors, it is only in the 'kibbutz' in Hefer Valley that greater use is made of them than in the neighbouring 'moshav'. In the Gilboa and Southern Districts the quotas are almost equal, while in Western Jezreel, where the settlers of the 'moshav', which is the oldest in the country, were for a long time opposed to mechanized cultivation, the quota of tractors is double that in the neighbouring 'kibbutz'. Indeed in respect of the use of tractors a veritable revolution has taken place in the 'moshavim'.

THE YOUNGER SETTLEMENTS

In this section we shall discuss eight settlements—four 'moshavim' and four 'kibbutzim'—situated, like the samples we have already dealt with, in four districts of the country, namely, Western Jezreel, Western Galilee, North Sharon and the Southern District.

The following table overleaf indicates:

(1) Irrigated fruit-growing is not yet developed in the younger settlements, and many have no orchards at all. Three settlements, of the four enumerated, which engage in fruit-growing cultivate abandoned orchards allocated to them when they occupied the land.

(2) In the majority of cases the area under irrigated fodder is larger in the 'moshav'. The 'moshav' in District A cultivates no irrigated fodder at all, but unirrigated fodder crops—two dunams per capita—supplemented by purchased feed, enable it to maintain the

TABLE 42

Comparison of some per capita units in younger settlements[1]

District	Settlement	Irrigated orchards per capita (dunams)	Irrigated fodder per capita (dunams)	No. of cattle per capita	No. of laying hens per capita	No. of draught animals in the whole settlement	Tractors per capita	No. of dunams cultivated per tractor
A	'Moshav'	—	—	1·60	23	23	0·026	533
	'Kibbutz'	—	0·06	0·42	12	3	0·035	869
B	'Moshav'	1·54	1·11	0·67	11	40	0·025	1,000
	'Kibbutz'	2·02	0·82	0·28	60	5	0·029	577
C	'Moshav'	—	0·66	0·47	1	38	0·006	1,153
	'Kibbutz'	0·15	0·78	0·28	11	6	0·024	1,157
D	'Moshav'	—	0·55	0·48	7	39	0·005	923
	'Kibbutz'	0·84	0·36	0·22	4	8	0·015	854

largest dairy herd in the list. The 'moshav' in District C also has less irrigated fodder than the neighbouring 'kibbutz', though its dairy is larger.

(3) In all the young 'moshavim' listed the dairy herd is substantially larger than in the 'kibbutzim' in the same district. Taking the herd in each 'kibbutz' as an index (= 100) the figures for the 'moshavim' are 325, 292, 325 and 370 respectively. The tendency in the younger 'moshavim' to maintain a larger dairy herd is noticeable from the third or fourth year of their existence.

(4) Poultry farming has not yet achieved optimal dimensions in the younger settlements. The figures in the foregoing table are not suitable for comparison, as in two districts the average number of laying hens is higher in the 'moshav' and in the remaining two in the 'kibbutz'.

(5) Mechanization. With the exception of District B, where the area cultivated per tractor is almost twice as large in the 'moshav' as in the 'kibbutz', the differences between the two types of settlement in this respect are not significant. The number of draught animals in the 'moshav' in Districts A and B is eight times as much as in the neighbouring 'kibbutz'. Elsewhere the number of draught animals is five or six times that in the 'kibbutzim' in the same areas.

INCOME

In our analysis of gross income in these sixteen settlements we shall use the same classification as formerly, namely older-established

[1] Based upon balance sheets of the settlements used in our sampling and a.m. *Statistical Bulletins of the Audit Union.*

and young settlements, 'moshavim' and 'kibbutzim'. The role of agricultural and non-agricultural branches in the local economy will also be discussed. The basic data are derived from the balance sheets of the settlements for 1951/52.

This section of our study is still based upon the selection of four 'moshavim' and four 'kibbutzim', of more or less the same age in the Gilboa, Western Jezreel, Hefer Valley and the Southern District.

TABLE 43

Per capita gross income in different types of settlements

District	Type of settlement	Gross income per capita (£I.)	Income from non-agricultural occupations per capita (£I.)
Older-established settlements			
A	'Moshav'	1210 (1657)	—
	'Kibbutz'	1279	124
B	'Moshav'	916 (1431)	—
	'Kibbutz'	925	91
C	'Moshav'	1195 (1360)	—
	'Kibbutz'	1328	1020
D	'Moshav'	1371 (2170)	—
	'Kibbutz'	1407	1017
Young settlements			
A	'Moshav'	739 (879)	—
	'Kibbutz'	933	299
B	'Moshav'	Information not available	—
	'Kibbutz'	1223	493
C	'Moshav'	Information not available	—
	'Kibbutz'	1211	281
D	'Moshav'	337 (383)	—
	'Kibbutz'	640	199

Two sets of figures are given in the foregoing table for the 'moshavim'. The first is based upon the method of calculation already utilized and includes skilled tradesmen, etc., who are permanent residents of the 'moshavim'; the second figure in brackets indicates the per capita income of a farming family.

In the 'moshavim' agriculture is the only source of income (for farmers). Practically every 'kibbutz', however, has developed some non-agricultural enterprise which supplements the income from farming. Both of the veteran 'kibbutzim' in Western and Eastern Jezreel, which have not developed industrial enterprises to any considerable extent, derive 91% of their income from farming operations. Non-agricultural branches include workshops, transport services and outside employment. The two 'kibbutzim' in the Hefer Valley and Southern District, respectively, are highly industrialized, no more than 23%–28% of their income being derived from agriculture. In each there is a large factory and various workshops, while one runs a rest-house.

In the fourth year of their existence these young settlements had already achieved a substantial level of income. The income of two of the more developed young 'kibbutzim' does not fall far short of that of older 'kibbutzim'. The four 'kibbutzim' chosen as representative in our sample, are based mainly upon agriculture, though non-agricultural enterprises bring in a higher income than in the older-established 'kibbutzim' in Western and Eastern Jezreel.

A very substantial proportion of the income of the young 'kibbutz' in the Gilboa District—approximately £I.40,000—came from transport services and outside employment of its members, only £I.10,000 was derived from industry and workshops.

In the District B income from industry and workshops accounts for 74% of the total income from non-agricultural enterprises.

In the remaining two 'kibbutzim' non-agricultural undertakings, mainly workshops and transport services, bring in 23% and 31% respectively of the gross income. In the younger 'kibbutzim' the emphasis has been upon the development of agriculture. The 'kibbutz' in District C, which has plenty of water, fertile soil and highly favourable topographical conditions, has already achieved a high income level. In addition to agriculture, this settlement is concentrating upon marine fishing. So far it has not embarked upon the development of industries and it is doubtful whether it will find it necessary to do so in the near future. The main field of the 'kibbutz' in the fourth district is the development of agriculture, which indeed continued to make steady progress in the year under review.

SUMMARY

The influence exerted by the creation of the State of Israel upon the agricultural economy of the country has been far-reaching and the changes induced in its structure and dimensions considerable.

Arab agriculture has shrunk mainly as a result of the abandonment of farm lands. The effendi class, at one time of much importance, has virtually disappeared, while the number of small and medium farms has increased. Means of production and marketing organization have improved, income has risen, accompanied by a concomitant improvement in the standard of living.

Prospects for further development of Arab farming appear favourable, particularly following the introduction, under the influence of Jewish agriculture, of improved methods of cultivation. So far this influence has not changed the social structure and the way of life of the Arab village to any perceptible degree. The persistent efforts of the Government of Israel to introduce even a limited measure of co-operation, in the most vital spheres of the village economy, have proved sterile. Even where the object of co-operation has been so essential a public utility as a waterworks, the attempt failed and the authorities were compelled to fall back upon the expedient of a limited company managed by a State official.

There is, of course, no prospect whatever of introducing forms of village organization embodying loftier social ideals, such as the 'kibbutz', the 'moshav shitufi' or the workers' co-operative settlement. The authorities, however, must persevere in their endeavours to place municipal organization upon a more satisfactory basis, and during a later phase must renew efforts to establish co-operative societies for the supply of water, credit facilities and agricultural machinery, and for the marketing of produce.

In the Jewish 'moshavot' urbanization is proceeding rapidly. Industrial expansion in these areas is only too often the criterion by which the capacity of local civic leaders is gauged. The process, there is little doubt, will continue, but it will not engulf the whole country. The tardy advance of industry in Israel must of course be regretted for it constitutes a standard whereby to measure progress towards economic self-sufficiency. Every effort must be made to stimulate the widest possible distribution of industry, in the villages, in the 'kibbutzim' and in the 'moshavim'. A 'moshava' with a number of factories need not aspire to urban status. It is only in two of the 'moshavot' discussed that industrial development has assumed major proportions, and here too farming is still carried on.

But the establishment of new industries is a slow process and meantime the economic importance of citriculture is becoming more and more patent. In those 'moshavot', which in the past were renowned for the quality of their plantations, a new 'back to the groves' movement would make sound economic sense.

In certain 'moshavot' viniculture was sufficiently developed in the past to warrant the construction of two large and well-equipped wine-presses. But vine-growing has stagnated. Now, in view of the high professional standards of growers and the good prospects of marketing Israeli wines abroad, a parallel 'back-to-the-vine' movement must be launched.

Citrus groves and vineyards can support a large population and even if leaders of the 'moshavot' should wish to encourage the construction of factories there is still adequate room for industries processing local raw materials. Cultivation of cotton and flax, of tomatoes for the manufacture of puree, of sugar-beet, of ground-nuts and tobacco opens up new vistas for the development of textile, sugar, oil, canning and cigarette factories. Such 'moshavot' could develop simultaneously as centres of industry and agriculture besides attracting a considerable population engaged in tertiary occupations.

The middle-class villages have made a valuable contribution towards agricultural development in Israel. They have grown in number, in size and in productive capacity. They have adapted to their own conditions and needs diverse forms of economic and social co-operation, and already play an important role in the agricultural community and economy.

The workers' settlements have wrought economic miracles since the establishment of the State. The ingathering of the exiles has ceased to be an abstract concept, it has taken on flesh and sinew and has become a vibrant reality spelling a new hope for the Jewish people and other peoples who have shared a similar lot. They have grown in numbers, in area, in means of production, at a pace never foreseen.

The foregoing study, based upon the sampling method, has enabled closer investigation of processes of growth, and has made possible analysis of the differences distinguishing various types of village organization. In addition to features shared in common by 'moshav' and 'kibbutz' in the same area, or two 'kibbutzim' situated each in different areas, we have indicated the divergencies, the inevitb le concomitant of different natural conditions.

The samples chosen have underlined the different paths along which the main types of settlement, the 'moshav' and the 'kibbutz', have advanced in the organization of their farms; they have stressed the fact that in the 'moshav' the farm-unit is larger per capita, as measured in terms of irrigated orchards, fodder, cattle and poultry.

But these data do not constitute a full and adequate picture of the farm economy. The 'moshav' is based almost entirely upon farming; in most of the 'kibbutzim' we find varying proportions of income derived from non-agricultural occupations and enterprises, from workshops and factories run independently by the settlements themselves, or regional undertakings under some other form of ownership. The means of production at the disposal of the farmer in the 'moshav' consist entirely of agricultural equipment; the member of

108

the 'kibbutz', however, has a stake in other means of production (besides sharing in the income coming from outside labour). This factor too must be reckoned with in analysing the structural, social and other differences distinguishing the 'moshav' and the 'kibbutz'. But both types of settlement march in one direction—towards the restoration of a national life, nourished by roots in the soil.

8

INDUSTRY AND HANDICRAFTS
IN THE VILLAGES

REFERENCE has already been made elsewhere to the fact that in the 'moshavim' only persons who are not settlers engage in non-agricultural occupations. The latter may be classified in two categories—salaried employees and independent workers who are paid by whoever has recourse to their services. The independent workers include truck drivers and various skilled artisans.

The 'kibbutzim', from a very early stage, developed various workshops such as carpentry shops, smithies, laundries, bakeries, etc., to meet local requirements. Frequently these small plants were expanded to supply the needs of neighbouring—and at times even more distant—villages, too. This was particularly the case when the 'kibbutz' had skilled artisans in some trade capable of producing high-quality workmanship. From these modest beginnings larger plants, including mechanical wood and iron working shops producing packing cases, crates, wagons, tools, etc., for the market, emerged. In the course of time canning factories, at first to process citrus, and later to can fruit and vegetables, were founded. Larger bakeries were established to supply bread for whole districts.

The collectivist school in Israel favouring the establishment of large 'kibbutzim' strongly supported the development of industries and handicrafts in the settlements (as well as the exploitation of all opportunities for outside employment), as a means of securing the largest membership economically possible. Members of the 'kibbutzim' engaged in public works, building, stevedoring in the ports, developed passenger and freight transport services in their neighbourhood, hired out tractors and other agricultural machinery.

Handicrafts and outside employment have come to occupy a place

110

of increasing importance in the collective economy, as can be seen from the table for the income of the 'kibbutzim' in 1937.[1]

TABLE 44

Sources of income in 'Kibbutzim' (%)

	Agriculture	Invest-ments	General works	Handi-crafts	Outside employ-ment
22 old established 'kibbutzim'	47·1	13·8	13·4	5·5	20·2
16 young 'kib-butzim'	24·1	17·0	10·2	12·3	36·4

Ten years later handicrafts and industries accounted for a substantial share of production, income and employment in the economy of a comparatively large number of 'kibbutzim'. In the 'Hakibbutz Hameuchad' settlements in 1936, 1,046 workers, or 18·3% of the total, were employed in such enterprises, which produced no less than 34% of the gross income.[2]

In the settlements of the 'Hakibbutz Ha'artzi', however, in the same period, only 368, or 12·7% of the total membership, were employed in such undertakings. By 1946 the share of gross income derived by 'Hakibbutz Ha'artzi' settlements from non-agricultural enterprises and outside work was 31·6%.[3]

Factories for the manufacture of fruit juices and jams established during the Second World War to supply British troops continued their operations after the return of peace. Six plants employing 167 workers were registered during the Census of Village Industries carried out in 1949. By the Census of 1954 the number had grown to 11, employing 327 workers.[4]

In 1947 all factories in this branch of industry—both privately and collectively owned—established a joint organization for marketing, the supply of raw materials, etc.

Other food industries include factories for fish preserves, the preparation of meats and sausage, olive presses, bakeries, biscuit factories, flour mills, lucerne- and hay-drying plants, etc. In 1949 there were 11 food factories employing 289 workers. By 1954 this figure had risen to 42 plants with 715 workers.

[1] Report of the Workers' Agricultural Organization to its Fifth Conference, 1939.

[2] *The Kibbutz in Figures*, Nos. 35, 36, 1947.

[3] Report of the Workers' Agricultural Organization to its Sixth Conference, 1949.

[4] Statistics on another plant in this category could not be obtained. Not included in this figure is one factory in the course of construction.

The food industry has been dealt with first because the plants process agricultural produce. The wood and metal industries, however, are of far greater importance in village industry. The list of metal products manufactured includes agricultural implements, machines, equipment and factory installations, irrigation fittings, water meters, electrical goods and fittings, etc. In 1949 there were 22 metal works employing 487 workers in the 'kibbutzim'.[1] By 1954 the number had risen to 150 plants employing 1,603 workers.[2]

In 1947 the metal works operated by the settlements together established the 'Meshavek' Company to market their products on a commission basis, to purchase raw materials, to co-ordinate production and to finance storage.

The following table gives a comprehensive picture of workers' settlements, handicrafts and industries.

TABLE 45

Handicrafts and industries in the workers' settlements
and the number of workers employed therein in
1949 and 1954[3]

Branch of industry	1949		1954	
	No. of plants	No. of workers	No. of plants	No. of workers
Metal	22	487	150	1603
Wood*	14	345	136	1180
Food	11	289	42	715
Stone and cement†	2	150	13	316
Chemicals and plastics‡	5	61	8	98
Leather (shoes)	2	10	4	88
Textiles and clothing	2	25	9	82
Printing and paper	3	24	6	61
Miscellaneous§	5	108	9	162
Total	66	1499	377	4305

* Manufacturing furniture, builders' joinery, crates, boxes, barrels, plywood, planks, boats, accessories, handles, clothes'-pegs, etc.
† Bricks, tiles, roofing tiles, insulating materials, betonade, quarries, plaster of Paris quarrying and grinding, etc. One plant was not included in the Census.
‡ Soap, paints, essential oils, vegetable extracts, plastic utensils and fittings, chalk products, etc.
§ Brushes, washing machines, rubber, wax, ceramics, diamond cutting, etc.

---

[1] Three metal works already in operation were not included in the Census.
[2] One other plant was being reorganized at the time of the Census.
[3] (a) For 1949—*Census of Industry and Handicrafts in the Villages,* Ministry of Agriculture, April 1950. (b) Figures for 1954—supplied by the Workers' Agricultural Organization.

With few exceptions where the settlement operates a single plant, it is intended to meet domestic requirements. Frequently this is also the case where two plants exist—a metal and carpentry shop, for example.

In the course of these five years the expansion of the 'kibbutz' metal and timber industries has been striking, the number of enterprises increasing seven- and almost ten-fold respectively. The number of persons employed in these two branches grew three-fold, indicating that a comparatively large number of small plants were established. It is noteworthy that every settlement founded in this period erected a metal and carpentry shop to meet domestic needs.

The large number of small workshops is reflected in the following table.

TABLE 46

Size of workshops in 1949 and 1954

Size of workshop or plant (on the basis of no. of workers employed)	1949		1954	
	No. of plants	%	No. of plants	%
Up to 10	20	30·3	292	77·3
11– 20	22	33·3	47	12·4
21– 30	13	19·7	15	4·0
31– 50	5	7·6	16	4·2
51– 70	3	4·6	—	—
71–100	1	1·5	3	0·8
101–150	2	3·0	4	1·1
150 and above	—	—	1	0·2
Total	66	100·0	378	100·0

Generally speaking the smaller plants employing less than ten workers are intended to meet domestic requirements. Prior to 1949 very few settlements operated workshops even for this purpose. Efforts to establish such plants which were actively supported by the State have been intensified in the past five years.

About 100 of these plants (approximately 25% of the total) worked for the market. Seven plants, each employing some 100 workers and producing plywood, insulating boards, bricks and conserves, are reckoned among the larger factories in their various branches. A number also produce for export.

With the exception of two factories situated in 'moshavim' and two other enterprises—a factory and garage—owned by regional associations of settlements, all workshops and factories in the workers' settlements are operated by 'kibbutzim'.

The settlements engaging in industrial development had no capital of their own and the investments were financed by bankers' loans.

This lack of investment capital, there is no doubt, constituted a limiting factor, retarding the pace of industrial development in the villages.

Another limiting factor in the collectives is the lack of manpower. Despite their desire to observe the principle of self-labour, industrial plants in the collectives are compelled to employ outside workers especially during the rush seasons. In the workshops operated for domestic needs, of course, no hired labour is employed.

The 'kibbutz' factories are in need of better machinery and equipment, more investment and working capital and labour. This last can be secured only by the absorption of new members.

NEW TRENDS IN
AGRICULTURAL SETTLEMENT

MASS SETTLEMENT

WHATEVER goal a policy of land-settlement is designed to achieve, whether it be quite simply the production of essential raw materials and food-stuffs, or whether it is also prompted by considerations of national security, culture, society or politics, the human factor outweighs all others.

In the history of most peoples settlement on the land has been a sluggish, protracted process, as elemental and heavy as the soil itself. The Jews constitute a notable exception, and from the earliest period of their history, their emancipation from foreign servitude and resettlement in their own country has proceeded at a more impatient pace. The ancient Israelites made a hurried departure from the Egyptian 'house of bondage', and though the return to Zion after the Babylonian exile was a more leisurely process, in the modern period haste and secrecy were recurrent characteristics of the Zionist colonization of Palestine. Under the Ottoman regime, it was only possible to embark upon some building project under the cover of darkness, and even as late as the concluding years of the Mandatory administration efforts to infiltrate into prohibited zones and border regions were perforce made clandestinely.

But even after the proclamation of the State of Israel, the need for haste remained, though secrecy could be discarded. Housing had to be constructed on an unprecedented scale for the hundreds of thousands of immigrants streaming through the newly-opened gates of the country. Methods of agricultural settlement had, of necessity, to be improvised to cope with the pressure of this mass influx. But though the prospective settlers, unlike the pioneers who had preceded them, could now make their entry openly, 'borne on eagle's wings'—to use a prophetic phrase that enjoyed wide vogue in those years of almost-messianic enthusiasm—they came without the lengthy vocational, cultural and social training that many of the immigrants of the Mandatory period had undergone.

The new immigrants came from every corner of the globe—from Central and Eastern Europe, from the British Commonwealth and the United States, from Latin America, from North Africa, India, Persia, Iraq, Yemen, Turkey, Bulgaria, Greece, Yugoslavia, Manchuria and Kurdistan. But irrespective of where they came from, they had hardly any notion of cultivation of the soil prior to their arrival in Israel. They were mostly unaccustomed to manual labour. The six years that have passed have shown tens of thousands of the newcomers to be diligent workers, who have already struck root in the soil. But there were also many who were less successful, who rejected landwork, and even some who left the country. The fault, perhaps, was not entirely theirs, for they lacked the necessary guidance and training. It is true, the veteran farmers did all they could, and more, to assist these new settlers, particularly in the initial period of settlement on the land and the development of their farms. But the number of these instructors ought to have been increased ten- and even a hundred-fold to ensure the success of their efforts. Moreover it must be borne in mind that the newcomers were introduced to tillage by way of the 'moshav ovdim'[1]—the workers' co-operative settlement—a form of rural organization calling for a considerable measure of mutual trust and mutual aid and presupposing a high standard of culture and education towards organized society.

PROBLEMS OF ADJUSTMENT

Both the settlers and the settlement authorities were confronted —and indeed are still confronted—by grave sociological problems, generated by the existence of distinct clans and communities among the newcomers, and often had to choose between the alternatives of establishing a village with settlers coming from a single country— or even province—abroad, or endeavouring to mingle two or more such communities in a single settlement. The obvious advantage of mixing the settlers as much as possible is considerably diminished by the danger of internal dissension, lack of mutual confidence and the development of a feeling of inferiority among the less advanced element, all of which may jeopardize the very existence of the settlement. On the other hand, the creation of settlements with settlers coming from a single, more or less homogeneous community may

[1] 'Moshav'—settlement (in Hebrew). 'Ovdim'—plural of 'oved'— worker (in Hebrew).

lead to the transplantation to Israel soil of diaspora ways and customs, of family rivalries and struggles for communal supremacy. There have been cases of entire settlements being broken up, as a result of such feuds and quarrels, many of the would-be settlers leaving the country, despite a knowledge of the conditions awaiting them in their countries of origin. The causes of people leaving their villages are many and varied; and may include personal and domestic reasons as well as the normal difficulties of self-adjustment to new conditions. Nor can we ignore the complaints of insufficient encouragement when it was most needed, inadequate vocational guidance and delays and deficiencies in the provision of tools and equipment. However, at every period of re-settlement in this country a certain proportion of the settlers have left the land. Not all the Bilu[1] pioneers or the founders of the first settlements remained where they originally settled. There have even been cases of entire villages being abandoned by the early settlers, and it should be recalled that in these cases the people were leaving something started by their own initiative and not by any outside agency.

REASONS SETTLERS LEAVING THE VILLAGES

However, if we recall that in all the settlements established by the new immigrants the proportion of those who decided to leave has never exceeded 10%, it is obvious that desertion of the new villages cannot be regarded as endemic. The part played by objective factors is indicated by the comparatively high percentage—15%—of people leaving the Negev settlements, where conditions are naturally harder, and in the hilly regions, where the proportion was 11%. In the more favourable districts, however, such as the central area of Israel, the proportion of settlers leaving has never exceeded 7·5% of the total.

There can be no question that this process causes harm both to those who leave and those who remain. At the same time it is noteworthy that all the farms rendered vacant during the year 1952/53[2] were taken up by new candidates in the course of the same year, and that in many cases the number of settlers was even increased. This,

[1] A group of Russian Jewish students who came to this country in the early 1880's with the purposes of engaging in agricultural labour. The name 'BILU' which they adopted derives from the initial letters of their motto 'Beth Jaacob lechu unelcha'—'O House of Jacob, come ye, and let us walk in the light of the Lord' (Isaiah ii. 5).

[2] The agricultural year in Israel is reckoned from 1 October to 30 September. The year 1952/53 means from 1 October 1952 to 30 September 1953.

of course, does not imply that these villages have already achieved a degree of stability and that some of the new, or even older settlers, are not liable to leave. The process of consolidation, of creation of a permanent settlement, is a prolonged one. How long it will take depends, among other things, upon local conditions. In certain districts natural conditions are unfavourable and do not allow for the development of farms capable of supporting the settlers, who must needs have recourse to outside employment.

A survey of immigrants' settlements conducted in spring, 1953, showed that there were fewer cases of settlers neglecting their holdings when they had eight dunams under irrigation than when they had only three or four dunams. Clearly, therefore, we cannot blame the new settler for his lack of success until we have established whether the natural conditions of the district were not too difficult, or whether there were not serious deficiencies in the planning of the settlement, in the supply of machinery and equipment, in professional guidance, in the balancing of production, in the organization of marketing and payment for produce sold, and the supply of credit for working capital and the like.

That such errors of omission and commission were liable to occur was obvious from the very outset, for the new villages were established in haste, under the impact of war and a mass influx of immigrants.

BORDER SETTLEMENTS

The signing of the Armistice Agreements with the neighbouring Arab states left Israel with a long and tenuous frontier, that winds often through wild and desolate regions, where danger constantly threatens. The favourable conditions for infiltration and border incidents create a special problem for the Israel defence forces. Before the last echoes of the war had died away it was obvious to Israel's planners that the more desolate areas in the north and in the Negev, in the Jordan Valley and the Hebron Hills and even along the eastern border of the central district of the country, the Samaria and Sharon Valleys, must all be settled with the least possible delay.

Thus a type of settlement, of which some slight experience had been gained in the past—the border villages—was resuscitated.

In the earlier stages, collective and co-operative groups founded by pioneering youth organizations and ex-servicemen were despatched to develop these areas. When, however, the number of prospective settlers of this category declined, the establishment of the border settlements was placed in the hands of new immigrants. In the first two years of the State no less than 111 settlements of this type—of which one-third were co-operative 'moshavim' and the remainder collective 'kibbutzim'[1]—were founded, comprising 4,500 families numbering 15,000 souls, occupying an area of approximately 300,000 dunams.[2]

In the three years which followed this first wave of settlement another 41 border settlements were established, bringing the total up to 154, excluding the older pre-war frontier settlements.

NEW TYPES OF SETTLEMENT

New conditions gave birth to two new types of settlement—the 'Moshav Olim' (immigrants' settlement) and the 'Kfar Avoda' (labour village). At the same time the establishment of new villages of the conventional types continued at an accelerated pace. Most of the 'kibbutzim'—which continued to absorb new settlers as they developed and progressed—underwent considerable expansion in these years. The 'moshavim'—co-operative villages—also accepted new settler-members who occupied reserve lands in the immediate vicinity, or—in some cases—when water was struck and new and more intensive methods of cultivation could be adopted, made room for immigrants in their midst. The expansion of existing villages upon these lines developed into a common method of settling newcomers on the land, mainly as a result of the availability of abandoned lands, which were placed at the disposal of new settlers.

Within this context mention must also be made of forms of temporary settlement, which were in the nature of an introduction to permanent settlement on the land.

Even in the early period of the mass influx of new immigrants the existing network of immigrant hostels was found inadequate, and camps to house the newcomers were established in various parts of the country. Those situated in established or already developed areas were converted into work camps, as the immigrant camps constituted not only a serious drain on the national exchequer but exerted an undesirable influence upon the inmates, who were for the most part unemployed. Thus it was essential to provide the newcomers with work, and to develop some other, albeit temporary, way of life for them.

THE TRANSITIONAL WORK CAMP ('MAABARA')

The creation of the 'Maabara' (transitional work camp) constituted a further stage in the temporary settlement of newcomers to Israel. These camps were planned primarily with a view to enabling the

---

[1] 'Kibbutz'—a collective settlement. The first collective settlement was founded in the Jordan Valley in 1909. Plural: 'Kibbutzim'.

[2] Dunam—a land-unit = 1,000 sq. metres = 0·1 hectare = 0·247 acre.

new immigrants to work for their living, while at the same time they facilitated the development of somewhat better housing conditions —larger tents in the first phase, aluminium and canvas huts later— the construction of paved paths between the tents or huts, and even the cultivation of small plots of vegetables. In 1951, at the close of the third year of the State of Israel, almost 100,000 persons were still accommodated in 'maabarot' in various parts of the country.

The residents of these camps are employed in the vicinity and constitute a reserve of candidates for the establishment of permanent settlements—of either a rural or urban character. Not infrequently they are integrated into a neighbouring established settlement by the construction of more stable housing on the site of the 'maabara' or in the immediate neighbourhood.

The temporary forms of settlement—immigrants' camps, labour camps and 'maabarot'—discussed in this chapter are all features of an extraordinary project of mass settlement on the land which has been a concomitant of the process of mass immigration.

'MOSHVEI OLIM' (IMMIGRANTS' SETTLEMENTS)

The 'Moshavim' Movement mobilized groups of its settler-members who visited the camps to explain to the newcomers the benefits they could derive from settling on the land. Minimal requirements of the candidates were formulated in respect of age, physical capacity, the organization of an adequate group of settlers from the community in question and the like. Towards the close of 1948 an immigrants' settlement was established on the land of the abandoned village of Akir, between Rehovot and Ekron. This beginning had an encouraging effect within the immigrant camps, and in the course of the year 1949/50 65 immigrant villages were established and 59 in the following year. The pressure created by the tens of thousands of newcomers, who filled the camps in these years, compelled a pace of settlement for which the various authorities were not prepared. It was necessary to plan 124 villages, to construct the dwellings, to supply them with water, to equip them, all in a period of twenty months. Thus it proved necessary to establish new villages at the rate of one every three or four days, under conditions of an acute shortage of equipment, and a lack of personnel for instruction and management. Small wonder that under such conditions serious errors of judgment occurred and that it became necessary to slow down the pace. In 1951/52 14 new settlements were founded and in 1952/53—12, bringing the total established in the course of four years up to 150, of which 108 are affiliated to the 'Moshavim' Movement,[1] and the remainder to

[1] Ami Assaf, *Individual Settlements in Israel* (in Hebrew), 1954, p. 195.

120

the 'Hapoel Hamizrachi',[1] 'Haoved Hazioni',[2] 'Poalei Agudat Israel'[3] and other organizations.

POPULATION

According to statistics published by the Audit Union for the Agricultural Co-operatives there were, at the beginning of 1953/54, 155 new immigrant villages with a population of 45,868, of whom 21,551 were workers, 21,171 children (under working age) and 3,146 parents and relatives.

We have reliable data for 1952/53[4] for the area of land cultivated and the various agricultural branches developed in the new immigrant settlements. In this year, it must be recalled, the oldest of these villages was no more than four years of age while there were many even younger. It must be appreciated accordingly that agricultural development was in its initial stages. Nevertheless these figures prove that important agricultural progress had already been made.

TABLE 47

Area under cultivation and main branches of farming (dunams)

Total area	461,522
Permanent holding	256,979
Temporary holding	204,543[5]
Total area cultivated	262,069[6]
Percentage of total area cultivated	57
Irrigated area	47,638
Percentage of cultivated area under irrigation	18·2
Field crops	181,520
Field crops as percentage of cultivated area	66·6[7]
Orchards	24,138
Orchards as percentage of cultivated area	8·8
Green fodder	12,030
Green fodder as percentage of cultivated area	4·4
Potatoes and vegetables	35,823
Potatoes and vegetables as percentage of cultivated area	13·1

[1] An orthodox group of labourers.

[2] A group of the progressive (non-socialist) Zionist party.

[3] Another orthodox group.

[4] *Statistical Bulletin of the Audit Union for Workers' Agricultural Co-operatives*, No. 28, August 1954.

[5] In 'temporary holding' in this case were areas the leasehold contracts on which were not yet signed.

[6] This is the physical area, whereas the cultivated area was larger, i.e. 272,352 dunams.

[7] From here and to the end of this table the percentage will be based upon the cultivated (not the physical) area.

The remaining area is either wooded or being prepared for cultivation.

By that year the new immigrants already cultivated wheat, barley, oats, maize and sorghum, peas and beans, fodder, as well as orchards of deciduous fruit, the majority of the latter abandoned by their former owners. It is of interest to note that 87% of the olive groves

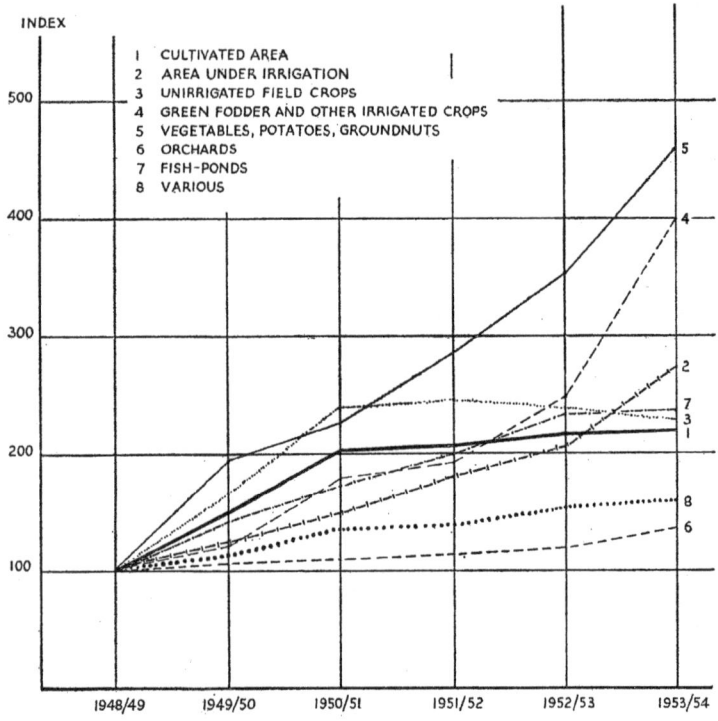

Fig. 1.—Area under Cultivation

of the 'moshavim' were being cultivated by immigrants. The newcomers cultivated in 1952/53 a total area of 24,138 dunams of orchards, of which 10,492 dunams were olive groves, 7,040 dunams—vineyards, 2,448 dunams—deciduous fruit, 2,312 dunams—citrus, 1,656 dunams—sub-tropicals, 150 dunams—bananas and 40 dunams—nurseries.

In the beginning of 1953/54 their livestock included 7,883 head of cattle, 8,279 sheep and goats, 73,900 laying hens, 2,614 draught animals, in addition to bee-hives, geese, turkeys, etc.

Their fields were being cultivated with the aid of 463 tractors, 72 combined harvesters, 63 presses and 131 lorries.

The contribution of these settlements to agricultural production was notable in the marketing of vegetables, which in 1952/53 were cultivated upon an area of 35,823 dunams. It should be borne in mind that in 1936 all the 'moshavim' and 'kibbutzim' in the country cultivated no more than 4,322 dunams of vegetables.

These new settlers came to this country practically destitute. Valuable assets were entrusted to them and their share in producing the nation's food and in farm production generally is already substantial.

DIFFICULTIES AND PROBLEMS

In view of the difficulties involved, and the achievements registered, the childhood ailments of these new villages, the abandonment of farms, the all-too-frequent election and re-election of local committees, which rose to power and fell with a rapidity reminiscent of the certain political regimes, can be forgiven and forgotten. And though the state of affairs in some of these villages was not entirely satisfactory in this initial phase, let us remember that no more than six years have elapsed, at most, since their foundation. A long road still lies ahead of them until their farms are fully developed, their social relations consolidated and their organization stabilized. But this task will prove less difficult, for the foundations have already been laid.

It is not enough to teach new settlers the elements of ploughing and seeding the land, the art of harnessing a mule and milking a cow. The farm is something more than a mere 'food factory', and there is always the danger, under conditions obtaining in this country, that if these villages are not inspired by national and political ideals, if their social basis is not stable and their organizational instruments not effective, if the minimum of co-operation is not achieved, then any crisis that farming, or the State or even the district in which the settlers live may experience, will be followed by the collapse of all that has been constructed and accomplished.

For this reason an onerous responsibility devolves upon the various agricultural, social and political organizations, to extend every possible assistance to these new settlements and to make their development and progress their major concern.

CANDIDATES FOR AGRICULTURAL SETTLEMENT ACT

This concern must express itself primarily in providing adequate instruction in farming for the settlers and supplying necessary farm

123

equipment and implements. But there may be members of the settlements, who are not only incapable of adjusting themselves to their new conditions, but constitute a disturbing and even destructive influence. Such persons, clearly, must be removed from the village. Some settlements have already learnt from bitter experience the damage that can be wrought by such elements in their midst and it has been found necessary to pass special legislation—the Candidates for Agricultural Settlement Act (1953)—under the provisions of which it is lawful 'to expel a candidate from an agricultural settlement in keeping with the provisions of this Act, if the demand for such expulsion is submitted before the candidate in question has been settled for a period of three years in the settlement'. The reasons for such a demand can be either economic or social. It is permissible to expel a candidate for settlement if the competent authority is convinced that the person in question (1) is persistently neglecting the cultivation of his holding without adequate reason; (2) is wilfully causing damage to the movable or immovable property which he has received from the settlement authority to enable him to settle on the land, or uses such property for uses other than settlement on the land, at the same time neglecting to use it for settlement on the land; (3) does not perform obligations imposed by the regulations of the settlement or obligations imposed upon him by the settlement in accordance with such regulations; (4) disturbs the peace of the settlement by acts of violence or threats of acts of violence.

Little recourse will probably be had to this Act. At the same time such legislation has been necessary to counter the influence of harmful elements introduced into new villages during the period of mass settlement, when selection of the candidates, prior to their joining the settlement group, was not possible.

Perhaps in the light of experience gained over recent years changes may be introduced into the methods in use. It might be advisable, for example, that the preliminary amelioration of the site for the future settlement should not be undertaken by the settlers themselves but by skilled workers under expert supervision. In so far as practicable, candidates for settlement in any area should be employed in that area. The settlers should be brought only after the dwellings and farm buildings have been erected, the plots and holdings parcellated, the roads laid out, the water supply installed, and communal institutions, such as the co-operative store, the clinic and the school, established. Livestock and an agricultural inventory should be handed over to new settlers shortly after their arrival. The construction of fences, the planting of woods and avenues and similar tasks could be performed by them, too. The main work of preparation, however, should be carried out by the settlement authorities and the new settler takes possession of a ready-built house. This method would probably be more economical. At the same time it removes basis for complaints by the settlers regarding the initial hardships of occupation of the land.

Of course use of this method is not necessarily suggested in all cases. But in respect of certain communities which over the past five years have proved incapable of overcoming the initial difficulties, and in the organization of the 'Town to Country' movement—which is dealt with later—this method offers a better prospect of success.

LABOUR VILLAGES

The manpower and funds at the disposal of the settlement authorities were insufficient in the period under review to meet the many and varied needs of the process of mass-settlement. To ensure the eventual establishment of fully equipped villages, various types of provisional settlements proved necessary. Thus, Labour Villages ('Kfarei Avodah') were established whose inhabitants were employed in land improvement works by the Jewish National Fund.[1] The work consisted mainly in clearing the land, terracing hillsides and in afforestation. The establishment of such villages was welcomed during the years when the immigrant camps were crowded, for it made possible the withdrawal of many newcomers from a life of corrosive idleness and their transfer to constructive employment. Twenty-six such labour villages were founded, numbering 10,000 souls in all. Half of these villages were established in the Jerusalem Hills and the remainder in Galilee and elsewhere. Each village comprised 100 families, housed in standard concrete or wooden dwellings, and sometimes in tents. The Jewish National Fund undertook to provide the settlers with a minimum of 240 days of work annually, while the right of the settlers to cultivate the land they improved was reserved. In the course of time most of these villages have developed into permanent settlements or surburban housing projects, in the Jerusalem district, and only nine have remained in their original form.

THE EXPANSION OF EXISTING SETTLEMENTS

The possibility of supplementing the holdings of many villages by placing tracts of abandoned lands at their disposal, the intensification of agricultural methods mainly by the extension of the irrigation network and finally the availability of new candidates for settlement

[1] J.N.F. was founded in 1902 for redemption of land on Jewish National Ownership.

on the land, combined to produce a new system of expanding established settlements by increasing the number of holdings. Some friction between veterans and new settlers must always be expected, especially in the co-operative settlement, where the farms of the former are in full production while those of the newcomers are still in the early stages of development. But the older 'moshavim' made strenuous efforts to facilitate the adjustment of the new immigrants they had absorbed to their new way of life. This process of expansion was equally important for the villages themselves. Of the 34 'moshavim' which were enlarged, in 16 the number of holdings was doubled, in 11 others the increase was 50%, while 7 were expanded by a quarter or a third. Of the 1,200 families settled in this way, 900 were new immigrants, and 300 ex-servicemen. In middle-class settlements 700 families were placed on the land in this fashion.

FROM TOWN TO COUNTRY

Efforts to settle urban families on the land have been common in many countries, their object being to reinforce the agricultural sector of the community, especially after periods of war. The difference between such movements elsewhere and their Israeli counterpart may be summed up in the term used for them overseas—'Back to the Land'. The title given to this movement in Israel—'From Town to Country'—adequately expresses its character under local conditions. In other countries farmers and peasants who for some reason had been drawn to the city but had not yet completely severed their domestic ties—nor even, often, their economic ties—with the land were resettled in the villages they had left.

In Israel the newcomer who is willing to settle on the land and to participate in the foundation of a new village is the main object of official attention in this sphere. The various stages of this process are already more or less a matter of routine. Settlement in an established village involves negotiations and special arrangements with the settlement authorities, though this method too, as described in the previous section, has already assumed substantial proportions. No attempt to enlist those already settled in the cities or large 'moshavot' was made until quite recently when the various settlement movements took the initiative. This time the call issued to the urban workers did not fall on deaf ears. Workers who for many years had cherished the hope that one day they would settle on the land and would see their children flourishing in the villages, close to nature, but had hesitated to take the decisive step, resolved to join the new movement.

Twelve new settlements were established as a result of the 'Town to Country' movement and a total of 6,000 former town-dwellers were transferred either to new or to existing villages. The division between 'kibbutzim' and 'moshavim' was roughly equal. This movement may develop into an important aspect of the work of re-settlement provided the necessary guidance is forthcoming and the authorities supply the equipment and funds required.

DIMENSIONS OF AGRICULTURAL SETTLEMENT

There were 259 villages—private, co-operative and collective—in Israel at the time of the foundation of the State. By the beginning of 1954 the number had risen to 586. This figure does not include work camps, farm-schools and farming estates, the number of which, too, has increased considerably in this period. In the course of five and a half years, accordingly, the number of villages has more than doubled: certainly a notable achievement.

The largest 'moshavot' were also affected by this process, but here it assumed the form of urbanization. This aspect will be dealt with separately elsewhere. The tables which follow reflect the growth in the three main types of settlement in this country—the private 'moshav', the workers' co-operative settlement and the 'kibbutz'.

Unfortunately figures for the earlier period of these villages are not available, but there is no question that they, too, have undergone rapid expansion in the first five years of the State of Israel. They have increased more than three-fold in number, attended by a commensurate increase in total and working population.

The number of villages of this type has increased three-fold during the period under review, while the population has grown to an even greater extent, as a result of the establishment of larger 'moshavim' and the expansion of existing 'moshavim', which formerly consisted of 30, 40 or 60 families.

The notable increase in the number of parents and relatives can be explained by the structure of the new communities settling on the land. The North African and Middle Eastern communities have

TABLE 48(a)

Increase in population

Private 'Moshavim'[1]	1947/48	1952/53
Number of villages	15	50
Population	(no figures available)	9,332
Number of workers	1,376	5,958
Children	(no figures available)	3,674

[1] The Department of Middle-class Settlements of the Jewish Agency.

127

TABLE 48(*b*)

Workers' co-operative settlement ('Moshvei ovdim')[1]

	1947/48	1953/54	Increase	Index 1947/48 =100
Number of settlements	72	250	178	347
Population	18,268	74,588	56,320	408
Working population	10,363	37,944	27,581	366
Local children	6,426	31,859	25,433	496
Outside children	368	158	−210	43
Israel Youth and Youth 'Aliyah'	280	89	−191	32
Parents and relatives	667	4,528	3,861	679

larger families. In the past, immigration into this country was made up to a very considerable extent of pioneers ('Chalutzim') who preceded the arrival of their families. In recent years, however, the practice has been to bring entire communities to this country and to avoid the breaking up of families. There are very few outside children, or even Youth 'Aliyah' wards in the 'moshavim'. In the past, too, their number was small.

TABLE 48(*c*)

'Kibbutzim'[1]

	1947/48	1953/54	Increase	Index 1947/48 =100
Number of settlements	145	216	71	149
Population	43,258	71,569	28,311	165
Working population	24,770	36,876	12,106	149
Permanent working population	19,570	31,540	11,970	161
Local children	12,751	22,998	10,247	180
Outside children	1,991	3,552	1,561	162
Israel Youth and Youth 'Aliyah'	2,694	5,925	3,231	220
Parents and relatives	970	2,218	1,248	229

In 1947/48 the number of collective settlements ('kibbutzim') was twice that of the co-operative settlements ('moshavim'), while the difference in population was even greater. This relation was radically changed as a result of the large influx of immigrants and the process of mass-settlement on the land. Mass-settlement led to a 150% increase in the number and the population of the 'kibbutzim' over a period of six years. The proportionate increase on the number and population of the 'moshavim', however, was 350% and 400% respectively. Nevertheless the 'kibbutzim' still retain their position of primacy in total and working population. The permanent working

[1] *Bulletins of the Audit Union for Workers' Agricultural Co-operatives* for the years 1947/48–1953/54.

population in the 'kibbutzim' has risen slightly more than the total working population, as children of the settlements have grown older and taken their place in the ranks of the workers. At the same time the number of temporary workers has declined as a result of the efforts of the collective settlements to reduce the proportions of hired labour employed.

Temporary workers in the 'kibbutzim' include candidates for full membership, skilled tradesmen employed on a wage basis, youth groups in training, etc.

In the 'kibbutzim' the increase in the number of local children, parents and relatives of settlers has lagged behind that of the 'moshavim', as settlers with large families did not join the 'kibbutzim'. As against this, however, the 'kibbutzim' have accommodated a large number—almost 4,000—of outside children—mainly children who

TABLE 49(a)

Increase in area

Private 'Moshavim'[1]	1947/48 (dunams)	1952/53 (dunams)	Increase (dunams)	Index 1947/48 = 100
Total area	60,670	135,505	74,835	223
Area permanently held	30,160	65,525	35,365	217
Area temporarily held	30,510	69,980	39,470	229
Under irrigation	13,900	35,165	21,265	253
Field crops	36,340	70,120	33,780	193
Orchards	3,690	13,100	9,410	355
Fodder	3,110	5,020	1,910	161
Potatoes and vegetables	7,080	17,065	9,985	241

go to the 'kibbutz' schools—as well as 6,000 wards of the Youth 'Aliyah' and Israel Youth Groups. These categories are not of any significance in the composition of the 'moshavim'.

The area under cultivation of the private 'moshavim' has more than doubled over the past five years, the three-and-a-half-fold increase in the area under orchards being specially noteworthy.

The increase in the aggregate land-holding of the 'moshavim'—489%—was greater proportionately than the growth in the number of villages and their population. It may be assumed that this area will enable the settlement of other families in both new and older-established villages.

The exceptionally high figure for the increase in lands temporarily cultivated merely indicates that contracts leasing these lands on a more permanent basis have not yet been signed between the owners of the land—the Jewish National Fund—and the farmers.

[1] Annual Reports of 'Bahan'—Audit Union of Middle-class Agricultural Co-operatives.

TABLE 49(b)

'Moshvei ovdim'	1947/48 (dunams)	1953/54 (dunams)	Increase (dunams)	Index 1947/48 = 100
Total area	190,495	932,394	741,899	489
Area permanently held	172,869	549,261	376,392	318
Area temporarily held	17,626	383,133	365,507	2,174
Cultivated area	151,710	673,420[1]	521,710	444
Cultivated area as percentage of total area	79·6	72·2		
Area under irrigation	38,397	132,698	94,301	346
Irrigated area as percentage of cultivated area	25·3	19·7		
Field crops	100,009	507,240*[2]	407,231	507
Field crops as percentage of total area	65·9	71·6		
Orchards	12,835	44,749[3]	31,914	358
Orchards as percentage of total area	8·5	6·6		
Green fodder	13,997	40,683	26,686	291
Green fodder as percentage of total area	9·2	5·7		
Potatoes and vegetables	8,039	67,608	59,569	841
Potatoes and vegetables as percentage of total area	5·3	9·54		

* From here the figures are for the area under cultivation, which in this year was 5% greater than the physical area of the land. For orchards, of course, this is the fixed area and the increase indicated is that of the physical area under fruit trees.

Potato- and vegetable-growing are the most flourishing branches in the co-operative villages and occupy an area of over 60,000 dunams, indicating an increase eight times the 1947/48 figure. For the newly recruited farmers this constitutes a very fine achievement.

Field crops are the second most popular branch, for it was only with the aid of extensive cultivation that it was possible to bring the 400,000 dunams of new land—in addition to the 100,000 dunams previously held—under the plough. In this branch there is still a

[1] This is the physical area in dunams. The cultivated area was then 708,174 dunams, or 5% more than the physical area. The area for the year 1947/48 is expressed in physical dunams.

[2] In addition 65,857 dunams were under green manure and fallow, and 20,923 dunams more were deep-ploughed for field crops and gardening.

[3] 8,463 dunams of land were being prepared for the planting of new orchards.

[4] The proportion of the area under potatoes and vegetables for the year 1953/54 was calculated on the basis of cultivated and not of the physical area.

certain amount of leeway to be made up, as in 1946/47 the 'moshavim' had 80% of their total holding under cultivation as compared with 70% in 1953/54.

The importance of field crops in relation to the total area under cultivation rose by 5·7% or by 407,000 dunams. The large and sudden increase in the area under fruit must be explained by the fact that abandoned orchards—mainly olive groves—were brought under cultivation by the 'moshavim'.

The irrigated area increased by 94,000 dunams, certainly no mean achievement. The value of new irrigation installations on this area was of a magnitude of £I.18 million. Prior to the Second World War, the total area under irrigation in Jewish farms—with the exception of citrus groves—was 40,000 dunams. This large expansion reflects a growth in the area under potatoes, fodder and orchards. In the next few years tens of thousands of dunams of land will be brought under irrigation, stimulating further expansion in the younger settlements.

In 1946/47 the land-holding of the kibbutzim was two and a half times that of the 'moshavim'. Since that year their holding has increased by more than one million dunams. In addition the 'kibbutzim' cultivate extensive areas of land upon a temporary basis, because, like the 'moshavim', they have not yet signed leasehold contracts for a longer period with the owners—the Jewish National Fund—and also because a number of 'kibbutzim' are temporarily cultivating lands in other parts of the country. This is also the case in regard to the 'moshavim', although not to the same extent. Fully 85% of the total land-holding of the 'kibbutzim' is under actual cultivation. They have increased their irrigated area by 58,000 dunams, most of it being under green fodder, potatoes and vegetables. Like the 'moshavim', the main increase in the area cultivated is sown with field crops. Here, however, the reason is the large area cultivated temporarily, mainly in the Southern District. This large increase in the area under field crops, which in 1954 covered 72% of the area cultivated, distorts the relative expansion of other branches of farming, which have been extended by tens of thousands of dunams. Taking the distribution of land among the various branches, in the 'kibbutzim' and the 'moshavim', respectively, as a criterion, the production of fodder, which constitutes the basis of dairy-farming, increased to an almost equal extent, both relatively and in physical area. The same can be said of field crops. In the growing of vegetables, however, the increase in area was far greater in the 'moshavim' —the latter extending the area under vegetables by some 50,000 dunams as compared with only 20,000 dunams in the 'kibbutzim'.

Some figures about other assets of the 'kibbutzim' will complete the picture. Complete statistics for the new buildings, such as stores,

TABLE 49(c)

'Kibbutzim'	1946/47 (dunams)	1953/54 (dunams)	Increase (dunams)	Index 1946/47 =100
Total area	468,622	1,630,775	1,162,153	348
Area permanently held	396,318	761,266	364,948	192
Area temporarily held	72,304	869,509	797,205	1,203
Total cultivated area	331,520	1,385,605[1]	1,054,085	418
Cultivated area as percentage of total area		85·0		
Area under irrigation	72,625	198,271	125,646	274
Irrigated area as percentage of cultivated area	15·0	14·3		
Field crops	227,197	998,457[2]	771,260	439
Field crops as percentage of cultivated area	68·5	72·1		
Orchards	22,127	62,600	40,473	283
Orchards as percentage of cultivated area	6·7	4·5		
Green fodder	23,902	69,291	45,389	291
Green fodder as percentage of cultivated area	7·2	5·0		
Potatoes and vegetables	11,476	31,977	20,501	278
Potatoes and vegetables as percentage of cultivated area	3·5	2·3		

cold-storage plants, creameries, residential quarters, culture halls, schools and the like, are unfortunately not yet available. Perforce we must rest content with the statistics for the increase in livestock and farm machinery and equipment.

In March 1952 a census of farm machinery was conducted in Israel. The results were as follows:[3]

Tractors	3,133
Grain harvesters	783
Green fodder harvesters	235
Motorized sprayers	897
Fodder presses	563
Drills	759
Miscellaneous	16,917

---

[1] Physical dunams. The cultivated area was 1,417,392 dunams, or 2·3% more than the physical area.

[2] In addition 247,298 dunams were under green manure and fallow and 27,754 dunams were deep-ploughed for field crops and gardening.

[3] 'Census of Agricultural Machines in March, 1952', *Statistical Information*, Vol. III, No. 7, 4 February 1953, p. 105.

TABLE 50
Livestock and farm machinery

	1947/48	1953/54	Increase	Index 1947/48 = 100
Cattle (head)				
'Kibbutzim'	12,073	24,963	12,890	207
'Moshvei ovdim'	11,644	27,960	16,316	240
Private 'moshavim'	1,390	2,600	1,210	187
Sheep (head)				
'Kibbutzim'	18,411	38,350	19,939	208
'Moshvei ovdim'	2,133	13,929	11,796	653
Private 'moshavim'	1,170	1,945	775	166
Poultry (laying birds)				
'Kibbutzim'	298,324	628,471	330,147	210
'Moshvei ovdim'	428,694	620,726	192,032	145
Private 'moshavim'	327,000*	290,000†		
Machinery				
'Kibbutzim'				
Tractors	385	1,856	1,471	482
Combined harvesters	168	541	373	322
Lorries	330	966	636	293
'Moshvei ovdim'				
Tractors	53	988	935	1,864
Combined harvesters	37	211	174	570
Lorries	52	411	359	790
Private 'moshavim'[1]				
Tractors	(no figures)	128		
Combined harvesters	(no figures)	30		
Lorries	(no figures)	55		

* 1951/52.   † 1952/53.

TABLE 51

Draught animals (head)	1947/48	1953/54	Increase	Index 1947/48 = 100
'Kibbutzim'	1,359	862	− 497	63
'Moshvei olim'	1,864	5,938	4,074	318
Private 'moshavim'	938	1,504	566	160

The above tables show that in both the 'kibbutzim' and the 'moshavim' the number of cattle have doubled. Flocks of sheep doubled in the 'kibbutzim', and increased more than six-fold in the 'moshavim'. This latter increase must be attributed mainly to the supply of goats to the immigrants' settlements, to provide milk for domestic consumption and for cheese-making. Sheep-breeding is

[1] Annual reports of the Audit Union for Workers' Agricultural Co-operatives and 'Bahan'—Audit Union for Middle-class Agricultural Co-operatives.

being expanded following the introduction of flocks into the Negev settlements.

Poultry-breeding expanded by 150% in the 'moshavim' and 210% in the 'kibbutzim', despite the shortage of feeding grains which made it necessary to decrease the number of birds. In private 'moshavim', where in the beginning of the period under review poultry-breeding underwent rapid expansion, such a decrease was indeed effected. The overall increase in the number of layers, despite efforts to restrict poultry-breeding, may be explained by the allocation of chickens —a quota of fifty per farm-unit—to the new settlements.

The sudden increase in the area suitable for field crops and the need to cultivate them with the utmost speed and intensity, made it necessary to equip the settlements with heavy machinery. In this period over 2,000 new tractors, some 600 combined harvesters and 1,000 lorries were acquired, besides a large number of smaller machines. The number of tractors in the 'kibbutzim' multiplied almost five-fold, while in the 'moshavim', where with the exception of the villages based upon field-crop farming they had been used previously only to a minor extent, and had averaged no more than one tractor to the village, the use of heavy machinery became widespread. In the 'moshavim' tractors, combines, lorries and the like are usually the property of the local Village Committee or Co-operative Society, although in 1953/54 out of 988 tractors, 395, or 40%, were privately owned by settlers; 27% of the total number of lorries in the 'moshavim' were also owned privately. These machines constitute an additional source of income for the owners. Only five combined harvesters were not owned by a Village Committee or Co-operative Society. Unlike tractors or lorries, combines are not ordinarily hired out. The competition of tractors is slowly making draught animals obsolete in the 'kibbutzim' and in 1953/54 the total number had been reduced to no more than 63% of the 1947/48 figure. A contrary process was registered in the 'moshavim', however, where the number of draught animals increased by 5,000 head, over a period of six years, the brisk demand encouraging continued imports from abroad.

CHANGES IN CULTIVATED AREA

We have gone into some detail regarding the expansion of agriculture as expressed in terms of land under cultivation, livestock, machinery, etc. Tables 52 and 53 constitute a summary in respect of the increase in area, production and value of production.

The national average underlines the striking progress registered by the workers' settlements. Thus while the national average increase

TABLE 52

Area under cultivation (1948/49–1953/54)[1] (thousands of dunams)

	1948/49		1949/50		1950/51	
	Dunams	Index	Dunams	Index	Dunams	Index
Cultivated area	1,650	100	2,480	150	3,350	203
Area under irrigation	290	100	350	121	440	152
Unirrigated field crops	1,066	100	1,775	167	2,550	239
Green fodder and other irrigated crops	65	100	79	121	118	181
Vegetables, potatoes, ground-nuts	69	100	133	193	157	228
Orchards	355	100	377	106	392	110
Fish-ponds	15	100	22	147	27	180
Various: smallholdings, nurseries, flower-growing, sugar-beet	80	100	94	118	106	133

	1951/52		1952/53		1953/54	
	Dunams	Index	Dunams	Index	Dunams	Index
Cultivated area	3,375	205	3,550	215	3,650	221
Area under irrigation	510	179	600	207	750	259
Unirrigated field crops	2,599	244	2,549	239	2,450	230
Green fodder and other irrigated crops	127	195	165	254	260	400
Vegetables, potatoes, ground-nuts	198	287	244	354	300	462
Orchards	410	115	433	122	475	134
Fish-ponds	30	200	35	233	35	233
Various: smallholdings, nurseries, flower-growing, sugar-beet	111	139	124	155	130	163

was over 200%, the 'moshavim' and 'kibbutzim' increased their area under cultivation more than four times. The fact that the area cultivated with field crops by the 'moshavim' and the 'kibbutzim' grew five-fold, as against a national average increase of little more than 200%, reflects the willingness of these settlements to undertake the cultivation of lands left fallow, even on a temporary basis.

The increase in the area under irrigation is practically identical with the increase in irrigated cultivation in the 'kibbutzim' and 'moshavim', both of which utilized the new irrigation facilities to cultivate more green fodder. A few thousand dunams of orchards and groves were planted by other settlements. The 'moshavim' and the 'kibbutzim' increased their vegetable, potato and ground-nut plots by 80,000 dunams, out of a total national increase of 150,000 dunams.

[1] *Statistical Monthly of Israel*, Part B—*Economic Statistics*, Vol. 5, No. 3, March 1954, p. 160.

The foregoing table reflects the steady growth, over six years, in the total area under cultivation and in all the various branches of farming, with slight fluctuations in the area under unirrigated field crops, which after four years of expansion have been followed by two years of contraction. Probably the decrease of 150,000 dunams registered between 1951/52 and 1953/54 is accounted for by an increase in various intensive branches of cultivation. This is borne out by the increase of 133,000 dunams in the area under irrigation in these two years. This would reflect the process of intensification, which is beginning to overtake the extensive areas brought under cultivation in the first four years of Israel's statehood. In the initial two years of the State the area under cultivation increased by no less than 103%, but in the subsequent three years the rate of increase was no more than 2%, 10% and 6%, respectively.

The most remarkable expansion was in vegetable-growing, followed by fodder and other irrigated crops. The growth in the area under fruit has lagged behind, but this can be explained by the large investment required for the development of orchards and groves. In the course of five years Israel's orchards—citrus, deciduous varieties, vineyards and banana groves—have grown by 120,000 dunams or an annual average of 24,000 dunams.

The value of agricultural production in the year 1952/53 was nominally £I.223 million, of which the Jewish farms accounted for 91% and non-Jewish farms for 9%. However, this figure cannot serve as a criterion of the increase in production in view of the rapid rise in prices following inflation of the currency. Some idea of the advance in monetary terms may, however, be obtained from the table on page 138.

While there is not necessarily any correlation between the increase in the area under cultivation and the value of farm production—as the aggregate of the latter is determined by the crops cultivated, the yield from year to year, fluctuations in price levels and the like—the close relation between the value of production and changes in the distribution of the various branches of farming is patent. Field crops and vegetables occupy first and second place respectively in the rise in the value of production, this increase being proportionate to the increase in the area under these crops. The ratio between the indices of the value of production between 1952/53 and 1949/50 was 140 (183 : 131) while the index for the expansion of the area under cultivation in this same period was 143 (215 : 150).

Thus it may be said that the increase in the area under cultivation under conditions of a given distribution of farm-branches was commensurate with the rise in the value of farm production.

The yields obtained over a period of years provide further evidence

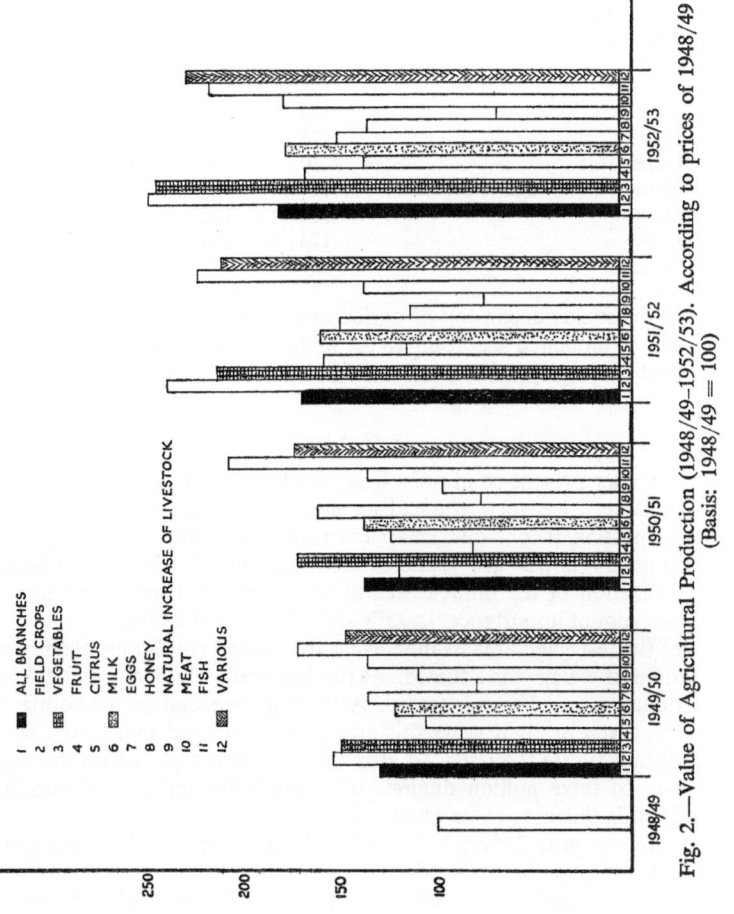

INDEX

Fig. 2.—Value of Agricultural Production (1948/49–1952/53). According to prices of 1948/49
(Basis: 1948/49 = 100)

1   ALL BRANCHES
2   FIELD CROPS
3   VEGETABLES
4   FRUIT
5   CITRUS
6   MILK
7   EGGS
8   HONEY
9   NATURAL INCREASE OF LIVESTOCK
10 MEAT
11 FISH
12 VARIOUS

TABLE 53

Value of agricultural production (1948/49–1952/53)
according to prices of 1948/49.[1] (1948/49 = 100)

	1949/50	1950/51	1951/52	1952/53
All branches	131	138	169	183
Field crops	155	119	241	250
Vegetables	151	172	213	245
Fruit	87	82	158	168
Citrus	108	123	116	137
Milk	122	137	160	177
Eggs	135	162	151	152
Honey	80	78	113	136
Natural increase of livestock	111	98	76	70
Meat	136	136	138	160
Fish	173	208	224	217
Various	147	174	213	230

of the steady expansion of agriculture in this country. These figures
will be tabulated according to the main branches of farming, namely,
field crops, vegetables, fruit, citrus, milk, eggs, meat and fish.

FIELD CROPS

Under present conditions it is possible to cultivate only one-fifth
of Israel's territory. Radical amelioration works may increase this
proportion to one-quarter or even one-third. But even if the most
sanguine hopes are realized the question of the most efficient
utilization of the limited area at the country's disposal remains of
paramount importance. Israel's agriculture must be intensive in view
of the restricted area available and the natural conditions of the arid
zone in which it is situated. At the beginning of 1955 some 800,000
dunams were being irrigated. Assuming even that in the course of
six, seven or eight years this area is doubled, and that in the more
distant future the irrigated area will aggregate two million dunams,
two to three million dunams will remain for unirrigated farming.
How is this area to be planned?

In the year 1952/53, 1,840,246 dunams were under unirrigated
cultivation as in Table 54.[2]

At least 80% of the entire area under field crops, including barley,
hay, fodder, sown pastures, legumes and oats, were intended for the
feeding of livestock. After deducting the area under green manure,
no more than 15% remains for the production of foodstuffs for

[1] *Statistical Monthly of Israel*, Part B, Vol. V, No. 5, May 1954, p. 323.
[2] *Statistical Abstract of Israel*, 1953/54, No. 5, pp. 56–57.

TABLE 54

Land under unirrigated crops (dunams)

	All farms	Jewish farms	% of all un-irrigated crops cultivated by Jews
Total	1,840,246	1,516,009	100·0
Wheat	346,912	195,649	12·9
Barley	827,488	677,103	44·7
Oats	19,599	19,579	1·3
Hay	376,868	376,410	24·7
Unirrigated fodder	7,305	6,635	0·4
Sown pasture	10,355	10,355	0·7
Ensilage	22,473	22,473	1·5
Green manure	82,446	82,446	5·5
Legumes	125,470	106,154	7·0
Flax	7,690	7,690	0·5
Sugar-beet	2,338	2,338	0·2
Peas (canning)	10,915	8,790	0·6
Various	387	387	—

human consumption. Thus it is clear that field crops serve the live-stock branches both in the production of grains and fodder. A controversy has developed recently in Israel over this question, in the course of which it has been proposed that a redistribution of crops be effected with a view to increasing the production of vegetable foodstuffs for direct human consumption at the expense of the live-stock branches, and to restrict, or at least halt, the expansion of dairy-farming. However, it would prove extremely difficult, if not impossible, to convince farmers of the advantages of producing cereals for the market instead of feed, which is converted into a more valuable form of live proteins such as milk, eggs and meat.

Barley is the predominant winter crop, both because it is the main concentrated feeding stuff for cattle and poultry, and because it produces a crop even in the Southern District, bordering on the Negev. This situation will continue as long as farmers are not convinced that some other variety of grain is more productive. Barley covers 45% of the area under field crops, followed by hay which accounts for 25%.

For many years Israeli farmers have been experimenting with new varieties of barley and other grains, in the hope of obtaining higher yields. It seems, however, that a lengthy period for the acclimatization of foreign varieties that appear suitable is still necessary. Efforts in this direction must continue in the experimental stations as well as in the fields until seeds suitable for local conditions are obtained. In the 1953/54 season it was found that foreign varieties suffered damage over extensive areas in the Southern District, though the local variety remained unaffected. While crop expectations of the foreign variety were considerably higher, it proved an excessive risk to allocate to it 90% of the area available. The local variety has proved that though its yield is moderate, the crop is nevertheless more certain, as it is fully adapted to local conditions.

Only Jewish farmers go in for the cultivation of fodder, ensilage and artificial pastures, which provide the basis for more intensive agriculture. Green manure, too, is confined to the Jewish sector. The cultivation of green manure is becoming more widespread from year to year, and despite the fact that it involves the loss of a year's crop, it has been found that the increased yield of the subsequent year is an adequate compensation. Flax, which is being cultivated in the Southern District as a fibre and oil-seed crop, is also confined to the Jewish farms. A factory to process the flax fibres has already been constructed.

Sugar-beet, at present grown to produce raw material for the distillation of alcohol, and to be used for the manufacture of sugar in a refinery under construction at Affule in the Jezreel Valley, is beginning to occupy substantial areas.

Yields

We shall now examine the yields of various crops obtained over a period of six years in Jewish farms.

Fluctuations in the *wheat* crop are not of major importance. They stem from vagaries of the seasons rather than from any change in the area sown. The average yield over the past six years has been 19,000 tons, and it is hoped to achieve an annual yield of 25,000 tons. The present yearly requirements of the country are 260,000–270,000 tons of wheat, not taking into account the possibility of an admixture of barley or maize meal. Local production, accordingly, can supply only 8%–9%, or at the very most 10%, of the total consumption. Thus, seeing that it is in any case necessary to supplement local production of foodstuffs by imports, the most feasible plan is to import wheat, as do Western European countries which suffer from a similar shortage, and to devote the restricted area at the farmers' disposal to the production of feeding grains, which are cheaper, but whose yields are more stable and the cultivation of which is more widely distributed. Perhaps in the more distant future, when the irrigation network covering the country will make possible the irrigated cultivation of two million dunams of land, it may prove practicable to test the growing of wheat in some crop rotation on land that has absorbed moisture after several years of irrigation.

TABLE 55

Yields during the period 1948/49–1952/53 (tons)[1]

	1948/49	1949/50	1950/51	1951/52	1952/53	1953/54
Wheat	16,500	19,700	10,000	22,000	21,000	23,800
Barley	17,000	28,600	25,000	85,000	58,000	77,600
Oats	900	1,400	1,200	3,500	2,000	1,400
Maize and sorghum	7,650	17,500	3,500	10,500	23,000	40,500
Legumes	1,600	1,500	650	5,350	2,500	5,550
Hay	40,500	53,000	55,000	120,000	110,000	126,000
Green fodder and ensilage	373,000	494,700	497,250	555,250	642,000	671,000
Ground-nuts	300	700	2,600	3,000	8,000	15,000
Other oil crops	550	2,650	600	1,750	1,250	2,100
Tobacco	—	35	105	230	150	150
Sugar-beet	—	—	—	4,500	8,150	6,750
Cucurbits	7,500	15,400	8,000	20,300	30,200	21,000
Flax	—	—	—	—	1,950	380
Straw	40,000	29,900	13,000	50,000	55,000	66,000
Miscellaneous (rye)	—	100	90	90	—	—

Very substantial progress has been made in *barley* yields. While here, too, seasonal influences are apparent, the above table shows that the crops of 1951/52 and 1952/53 were five and three and a half times as large, respectively, as that of 1948/49, indicating the preference of local farmers for barley over other cereals.

The cultivation of *sorghum and maize*, too, is undergoing constant expansion. Maize is a very important crop for this country in view of its high nutritive value for both man and beast. With auxiliary irrigation high yields can be obtained from this summer crop—five times higher, indeed, than wheat. Irrigation can ensure the stability of the average yield. Excellent results have been obtained both in the United States and in this country from the planting of cross-bred maize, and an increase in the cultivation of this crop can provide a firm basis for poultry-farming, the expansion of which is being restricted owing to shortage of local feeding stuffs. There is a prospect that in the more distant future, following an extension of irrigation, the surplus of maize may enable the admixture of maize flour—to an

[1] (1) For the year 1948/49—*Statistical Bulletin of Israel*, Vol. I, No. 6, March–April 1950, p. 402. (2) For the years 1949/50 and 1950/51—*Statistical Bulletin of Israel*, Vol. III, No. 3, June–September 1952, p. 90. (3) For the years 1951/52 and 1952/53—*Statistical Abstract of Israel*, 1953/54, p. 74. (4) Preliminary report of the Bureau of Statistics, June 1955.

extent of 10%–15%—to wheaten flour, thereby alleviating to some extent the problem of importing wheat for bread.

The cultivation of *legumes* is far less than it should be in view of the general lack of proteins for both man and beast in this country. But it is a difficult crop and is not very popular with farmers. The introduction of legumes into crop-rotation systems is extremely desirable in view of their property of enriching the soil, through the biological process in the course of which nitrogen is released about the roots.

The yield of *hay* has grown three-fold over the past five years. The crop of *green fodder and ensilage*, too, is increasing by 100,000 tons annually to meet the need for a hygienic, juicy and vitamin-rich feeding stuff to maintain health standards of stock and to increase milk yields.

The quantities of *ground-nuts* previously cultivated in this country were not sufficient even to meet the local demand for peanuts. Exceptionally rich in oil, ground-nuts, grown as a summer crop, produce a very high yield under irrigation. Indeed by growing ground-nuts upon some scores of thousands of dunams, the country's problem of raw materials for the margarine industry can be solved. The failure of the British ground-nut project in Kenya must be attributed to the excessively large scale upon which the project was conducted. In this country the growing of ground-nuts was launched experimentally at the beginning of the Second World War, and in the course of the years valuable experience in its cultivation has been gained. By the gradual expansion of the area cultivated the danger of a major setback has been obviated, and today, already, important results have been achieved. The yield, which in 1948/49 was no more than 300 tons, rose to 8,000 tons after five years, and it should be possible, within the space of a few years, to achieve an annual crop of 40,000 tons, and even more. There is a favourable market for Israel ground-nuts abroad as the nut is of good quality. Transport over long distances presents no problem as ground-nuts keep well. Ground-nuts are also highly popular in this country, especially among the Eastern communities. The crop can be exchanged partly for oil, which can also be produced in this country. It is highly desirable to reduce the price paid for ground-nuts and thereby to ensure its wider use in view of its high nutritive value as a source of oil and protein.

Vegetable-growing is rapidly approaching saturation point and for this reason its demands upon soil, water and labour resources will not increase to any considerable extent. Special importance consequently attaches to the introduction of sugar-beet as a winter crop and ground-nuts as a summer crop in the crop rotation.

Other oil seeds—*safflower and sunflower*—occupy second place after the cultivation of ground-nuts. An increase in the area under these crops is also advisable in order to diversify the variety of oil seeds produced in this country and to reduce dependence upon a single crop for the production of oil.

Tobacco. Arab peasants have been cultivating Oriental varieties of tobacco in this country over a long period. Jewish farmers made an unsuccessful attempt to grow tobacco as far back as 1924/25, the main reasons for their failure being inadequate cultivation and lack of experience (similar reasons were probably responsible for such debacles in other countries). The figures in the above table refer exclusively to the Jewish farms, for the yields of the Jewish and Arab farms have varied considerably and a special chapter will later be devoted to the Arab sector of farming. The relative importance of tobacco in Israel's agricultural economy is reflected by the fact that on Jewish farms the yield over the years 1949/50–1952/53 was 35, 105, 230 and 150 tons respectively, while on Arab farms in the same years it was 1,450, 1,800, 2,450 and 1,650 tons. In 1948/49 the Arab farms produced 600 tons of tobacco while the Jewish farmers did not engage in the cultivation of this crop at all.

Tobacco is accordingly an important crop for Arab farmers. The settlement authorities are endeavouring to extend the cultivation of tobacco in the hills and on the mountain slopes, mainly in the Jerusalem Corridor.

The trade in raw tobacco in practically all countries is conducted on a primitive and unorganized basis. In the absence of better relations between the manufacturers and the growers the prospects of this crop, based as it is upon a moderate standard of living, are uncertain.

For some years experimental cultivation of Virginia tobacco, which commands a higher price, has been continuing. The Government and the settlement authorities have jointly established the 'Alei Tabak' Company[1] to deal with the organizational and economic aspects of the cultivation of this crop. There are prospects that not only will Israel be able to supply its own consumption but that a certain surplus will remain for export, providing the marketing of the crop is properly organized under the supervision of the Government.

Sugar-beet. The experimental growing of sugar-beet was launched at the beginning of the First World War and continued intermittently thereafter. These tests were renewed during the Second World War and have continued to the present day. The shortage of foodstuffs in periods of emergency and the absence of competition have

---

[1] 'Tobacco Leaf Co.'

provided a new stimulus for the local production of foodstuffs and even for the development of sugar, oil and other industries. The pro-longed period of experimental cultivation was fruitful of results and proved that the construction of a refinery to process the local crop would be economically justified. The commercial cultivation of sugar-beet was accordingly launched and the crop, which in 1951/52 totalled 4,500 tons, rose to 8,150 tons in the following year. A yield of 80,000 tons, the quantity that will be required annually by the first sugar factory in the country at present being constructed in Affule, can now be planned. Once sufficient experience is accumulated in the cultivation and the processing of the crop two or three other sugar factories may be established elsewhere. Meantime sugar-beet is being utilized for the production of alcohol. Sugar-beet will be introduced as a winter crop in the crop-rotation system.

Flax. During the thirties attempts were made to grow flax for the production of linseed oil. However, owing to the impatience of the oil producers and lack of encouragement, experiments in this field were discontinued for a period of twenty years. Experiments were renewed some time ago, with excellent results in the Southern District. In 1952/53 the crop, which totalled 2,000 tons, was processed in a small pilot plant established in the Southern District.

Present attempts to grow flax in this country and to process the crop in the small factory established, though still in the experimental stage, may well constitute the nucleus of a fibre industry using locally grown raw materials.

The growing of *cotton*, and of *agava* producing a coarser fibre, has been launched in the north and the south of the country, respectively. While it is still too early to draw any definite conclusions from this short period of experimental cultivation, they exemplify efforts to en-sure a domestic supply of raw materials for Israel's textile industry.

Cucurbits. Water- and sugar-melons are an important item of diet for a short period during the summer in Israel, filling a gap between the grape and citrus seasons. The demand for both varieties is con-stantly increasing and over the period of five years since 1948/49 the yield has grown four-fold.

Straw. The quantity of straw produced—depending upon the dimensions of the cereal crop—which totals 50,000 tons, does not meet the need of the farms for stable bedding, the preparation of organic manure, etc.

It is highly probable that the area under field crops will not be extended in the near future. Agricultural planners and farmers will concentrate upon raising yields and achieving greater equilibrium between the various crops, with a view to increasing the production of feed for the livestock branches—dairying and poultry-breeding—

and of industrial crops for the local production of oil, sugar and textiles.

Vegetables and potatoes. The cultivation of vegetables and potatoes was at one time not popular with Israeli farmers owing to the large amount of manual labour required, the constant vigilance called for and the difficulties of marketing a perishable crop. Another problem was the sharp and frequent price fluctuations. The production of vegetables for the market commenced under pressure of the political disturbances during 1936, when the Jewish market in this country was cut off from Arab growers, both in Palestine and the neighbouring countries, who had been major suppliers of foodstuffs.

Only twenty years ago local farmers did not know how to grow potatoes. In the course of time it was found that an essential precondition for successful cultivation was the import of seed-potatoes from northern countries. A number of varieties were tested, until the Up-to-Date potato was found to be most suitable for local conditions. Subsequently a number of other varieties were also acclimatized.

The rapid increase in the area under potatoes, vegetables and ground-nuts is reflected in the following table.[1]

TABLE 56

Areas under vegetables and potatoes (thousands of dunams)

Year	1948/49	1949/50	1950/51	1951/52	1952/53	1953/54
Dunams	69	133	157	198	244	269
Index	100	192	228	290	354	390

The main vegetable-growing area is the Coastal Plain. The new settlements, founded since the creation of the State of Israel, have made an important contribution towards the development of vegetable-growing and last year cultivated no less than 48% of the entire area under this crop.

The total yield of vegetables and potatoes increased as follows in the period under review (in tons):[2]

TABLE 57

Yields of vegetables and potatoes (tons)

	1948/49	1949/50	1950/51	1951/52	1952/53	1953/54
Total	96,500	135,500	157,500	202,000	235,000	258,000
Potatoes	25,400	34,500	36,500	46,000	55,000	78,000
Other vegetables	71,100	101,000	121,000	156,000	180,000	180,000

---

[1] *Statistical Abstract of Israel*, 1953/54, No. 5, p. 5.
[2] See Note 1, p. 141.

Jewish farmers produce 94% of the country's gross potato and vegetable crop, but it must be borne in mind that the area under irrigation in the non-Jewish sector is still small and on Arab farms vegetables are grown without irrigation.

The largest area of land is utilized for the production of tomatoes, followed by potatoes, ground-nuts, cucumbers, onions, carrots, cauliflower, cabbage, beetroot, peppers and spring onions, etc. (in that order).

Not very long ago vegetables and potatoes reached the Israeli

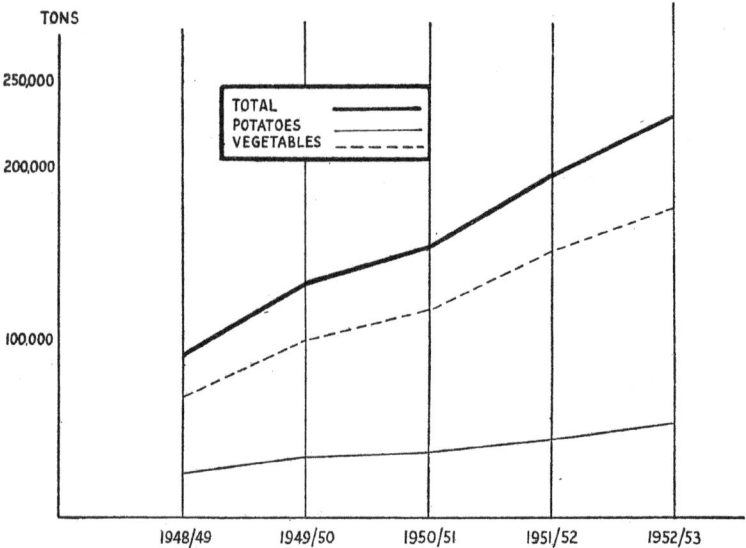

Fig. 3.—Total Yield of Vegetables and Potatoes in the Years 1948/49–1952/53

consumers' table from abroad, or were supplied by Arab farmers. Since 1953, however, practically all the country's consumption of vegetables is produced locally, while quantities of tomato puree are now being exported. Small quantities of potatoes are still brought from overseas during periods of seasonal shortage, but this, too, will soon be discontinued. The manufacture of canned vegetables and concentrates, and the improvement of local storage and refrigeration facilities, will help to bridge seasonal gaps in the supply of fresh vegetables to domestic markets.

The marketing of vegetables is beset by more problems than any other kind of farm produce, and much effort remains to be invested before the supply of vegetables to the cities is fully rationalized.

146

To meet local consumption farmers must increase production of certain varieties of vegetables, especially of potatoes in the spring and autumn, of carrots in spring to ensure supplies in the summer and of tomatoes in summer and winter.

The cultivation of potatoes upon the basis of prior agreement with the growers was successfully carried out in 1954, and large quantities were cold-stored, thereby ensuring regular supplies. Tests to establish the feasibility of exporting carrots are at present being made. With adequate cold-storage facilities it is also possible to ensure an unbroken supply of carrots from domestic production.

It is more difficult to organize a regular supply of tomatoes because of their perishable nature. The main season for tomatoes is spring, but the area planted in winter in the Jordan and Beth Shean Valleys can be increased. For the growing of tomatoes in summer, however, expert knowledge of cultivation and the acclimatization of suitable varieties not subject to the 'fusarium' disease is necessary. In recent years the cultivation of tomatoes for the production of puree has been begun.

Not long ago onions were not grown at all in the Jewish villages and only two years ago large quantities were still imported. Now local production can meet consumption fully provided onions are also cultivated in winter and a larger area is sown with the Egyptian and Cyprian varieties, which can be stored for longer periods.

Gratifying progress has also been made in the cultivation of other vegetables.

In a report submitted to the Conference of Vegetable Growers in the summer of 1952, the growers set forth the conditions favouring the expansion of this branch of farming, after the establishment of the State, as follows:[1]

(a) The vast growth in the population of the country and the increased demand for fresh vegetables;

(b) The keen interest evinced by the 'Yishuv'[2] and the various Government departments in the development of vegetable-growing in order to meet the consumption of the growing population of the State;

(c) The absence of outside competition in the market;

(d) Better cultivation, organization and agro-technical methods, and invaluable experience gained over a period of years;

(e) The establishment of new settlements.

New conditions and a more favourable atmosphere generally has

[1] Vegetable Growers' Association. Report submitted to the Sixth Conference, May 1952.

[2] The Jewish population (in Hebrew).

opened up excellent prospects for the expansion and development of this branch of agriculture.

FRUIT-GROWING

Fruit-growing was the only branch of Israel's agriculture in which no increased production was registered during the first three years of the State. It is true that the rehabilitation of orchards, which had been neglected or had suffered during the period of hostilities, was begun. Plans were also drafted for the planting of new orchards and substantial progress has since been made. In the past three years, however, the larger quantities of fruit marketed have been palpably felt by consumers, as is reflected in the table that follows:[1]

TABLE 58

Fruit yields during the period 1948/49–1953/54 (tons)

Variety	1948/49	1949/50	1950/51	1951/52	1952/53	1953/54
All varieties	28,900	26,100	24,330	36,930	41,300	55,300
Table grapes	9,000	7,750	5,600	8,800	10,100	14,400
Wine grapes	7,200	7,400	6,250	8,000	9,500	12,600
Olives	3,800	1,400	950	3,050	3,400	5,200
Bananas	3,500	2,000	5,680	8,180	10,900	11,300
Deciduous fruits	4,700	5,650	3,250	6,100	5,200	9,300
Miscellaneous	700	1,900	2,600	2,800	2,200	2,500

In the year 1949/50 the Ministry of Agriculture prepared blueprints for the development of fruit-growing in the ensuing four-year period as part of the overall plan for agricultural development. The plan was based upon an extensive survey of conditions in various parts of the country. The recommendations incorporated in the plan were based upon the estimated demand for fruit, the stock of seedlings, the expert personnel available and general conditions. The plan envisaged the planting of 75,000 dunams of new orchards and vineyards, excluding citrus. In effect in the period—1949/50–1952/53— 54,000 dunams of new orchards were planted. The plans were implemented, accordingly, to an extent of 72%. During these years a cartographic survey was conducted with a view to establishing the natural borders for each variety of fruit. The optimum area under orchards was placed at 504,000 dunams throughout the country, excluding citrus groves. The list is headed by unirrigated olive groves covering an area of 131,000 dunams, followed by vineyards (table grapes)—

[1] *Statistical Monthly for Israel*, Vol. V, No. 5, May 1954, and other bulletins of the Central Bureau of Statistics and Economic Research. For the year 1953/54—Preliminary report of the Bureau of Statistics, June 1955.

111,000 dunams and (wine grapes)—57,000 dunams; plums—37,000 dunams; apples and pears—29,000 dunams; bananas—23,000 dunams; irrigated olive groves—16,000 dunams and smaller areas under apricots, sub-tropical fruits, figs, almonds, dates, quinces and peaches. In the course of time the bounds set by this plan for the diverse varieties may change; meanwhile new planting is being carried out in accordance with its provisions. By the end of 1952/53 294,235 dunams were under fruit trees in the country (excluding citrus), leaving 210,000 dunams to be planted in the future. The task is by no means an easy one, for in addition to the immense investment involved—totalling tens of millions of Israeli pounds, of which 25% is required in foreign currency—extensive preliminary work is required in preparing the soil and seedlings (especially of sub-tropical varieties), installation of irrigation, the training of experts and the study of local and overseas markets. The rate of planting was as follows: 1949/50—3,000 dunams; 1950/51—14,000 dunams; 1951/52—17,000 dunams and 1952/53—20,000 dunams. More than half of the new area has been planted *with wine and table grapes*.

The new planting is being carried out by the 'kibbutzim' and by both the older and the new 'moshavim'. Private investors, local and foreign, are also interested, their orchards being concentrated mainly in the southern part of the country.

The *banana*, the cultivation of which was originally mainly confined to the Upper and Lower Jordan Valley, now covers extensive areas in the Coastal Plain, where, despite the danger of frost, conditions are favourable for banana-growing. There is a steady demand for bananas in this country, while efforts to develop export markets have proved successful. Plans drawn up for the development of banana groves are expected to be fully realized within five or six years, and if hopes entertained regarding export prospects prove justified, new plantings are expected to exceed the bounds set by the Ministry of Agriculture's blueprint.

From time immemorial the *olive* has been a native of this country and even now, despite the destruction of hundreds of thousands of dunams of olive groves, it is the most cultivated branch of horticulture. In the main the olive is utilized for oil production, and for this reason indeed it was not favoured by Jewish farmers, as the low prices for olive oil did not justify the high costs of hand-picking the fruit. Jewish growers, accordingly, have preferred the larger, pickling varieties. These groves are irrigated—unlike the oil olive—the crop possessing the dual advantage of high yields and good prices. Were the rehabilitation of existing olive groves by the grafting of better varieties, irrigation and improved methods of cultivation, possible, the yield might be increased from the present 12,000 tons in Jewish and Arab farms to 60,000–70,000 tons. New irrigated olive groves are situated mainly in the Beth Shean Valley and the Southern District closer to the Negev. An olive expert of international repute who has visited this country has recommended that olive-growing be fostered especially with a view to producing an Israeli variety of pickling olive.

Apples, which are highly favoured by local consumers, are grown in the hilly regions, the Western Jezreel Valley and the Hulah Valley. Even after taking into account the restricted area suitable for its cultivation, apple orchards can be extended to cover a total area of 30,000 dunams.

Though the *date* was indigenous to this country in the ancient period, the fruit of the local groves was found to be valueless and foreign varieties were introduced for the production of fresh dates and for drying. The development of high-quality dates is a prolonged process and, up to the present, date groves occupy only some hundreds of dunams. Promising attempts to cultivate the date palm have been made in the Jordan and Beth Shean Valleys as well as in the Negev down to Elath, and there appear to be good prospects to increase the area of groves up to a limit of 10,000 dunams.

Plums are the most important stone-fruit grown in this country. Only a small area is under *apricots*, while the beginnings of *peach* cultivation are meagre. The *cherry* has not yet aroused sufficient interest among growers. Provided growers plant new orchards in keeping with the provisions of the Government's plans, 50,000 dunams will eventually be cultivated with stone-fruits, about three times the present area.

Important experimental work has been conducted with a view to acclimatizing a number of varieties of sub-tropical fruits, but growers have concentrated principally upon the *avocado* and the *guava*, for both of which there is a good market. Of the two the guava is the greater favourite among growers. Plans for planting new orchards include 10,000 dunams under sub-tropical varieties—ten to twelve times the present figure—but it will be necessary to diversify the plantations and to introduce other varieties, such as the *persimmon*, the *loquat*, the *mango* and the *pineapple*. The sub-tropical fruits mentioned in this context are all newcomers to this country and require a lengthy period of acclimatization. Of course, the olive, the pomegranate, the date, the fig, the sycamore, the carob, all of which are sub-tropicals, are indigenous, while the citrus and the banana must also be regarded almost as natives.

Fruit-growing is an expensive branch of agriculture calling for substantial investments in capital and labour over a period of many years, before bringing in any return. For this reason, indeed, diversification is even more important here than in other branches of agriculture. A higher degree of professional skill and judgment is also required. The area under fruit will be doubled in the next ten to twelve years, preference being given to high-quality early and late varieties, to ensure a regular supply for local consumption and for export.

CITRUS

In 1947 there were 255,089 dunams of citrus groves in this country, of which 134,567 dunams (123,000 dunams in full bearing) were owned by Arabs, and 120,522 dunams (108,828 dunams in full bearing) were owned by Jews. The groves not bearing fruit had been neglected during World War II when exports were brought to a standstill.

Citrus-growing suffered more than any other branch of the agricultural economy from hostilities during the War of Liberation. The destruction, over a large area, of irrigation facilities proved the death-blow for many groves, which had already suffered from inadequate cultivation in previous years. In view of the extent of sabotage and deliberate incendiarism during the War there is little cause for surprise that only about half of the area formerly under citrus survived (taking into account the 24,000 dunams which have remained outside Israeli territory).

Of the 208,000 dunams of groves which bore fruit in 1942 and which were situated within the borders of Israel, 126,360 dunams were included in the census of citrus groves carried out in 1950/51. Of this area 28,571 dunams were cultivated by the Custodian of Absentee Property. In 1952/53, following considerable work of rehabilitation of neglected groves, and new plantings, the area under citrus was 142,000 dunams.

In 1947/48 11·5 million cases of fruit were produced, of which 9·9 million cases were exported. In 1948/49 the yield was 6·3 million cases, of which 4·2 million cases were exported. In 1953/54 production had risen again to 11·6 million cases, of which 8·1 million cases were exported, mainly as a result of rehabilitation work over a period of two or three years.

The following are the figures for the production, marketing, export and processing of citrus over the past five years.

Gross production	1,460,000 tons	100·0%
Export	825,000 ,,	56·6%
For manufacture of concentrates and juices	395,000 ,,	27·0%
Marketed locally	240,000 ,,	16·4%
Foreign currency income	84,000,000 dollars	

151

The Government of Israel has encouraged the rehabilitation of citrus-growing and has prepared detailed plans for this purpose with the aid of a staff of expert instructors. Government aid for the industry has included importation of specialized equipment—financed by the American Loan—to a value of $8,310,000, loans to growers to an amount of £I.2 million, the payment of premiums and finally allowing growers a rate of exchange of £I.1·8 to the dollar for income from exports.

Citrus, like other branches of agriculture, is subject to the vagaries of the seasons. The 1951/52 season was one of the worst growers ever experienced, as a result of unfavourable climatic conditions. Abundant rains, which were welcomed by other branches of farming, interfered with the picking of the fruit and reduced the exportable crop by half a million cases as compared with the previous year. Indeed the percentage of the crop exported was only 44% of the gross yield in that year as compared with an average of 56·6% over a period of five years. The year—1953/54—was an excellent season, for in addition to the increased quantity exported growers benefited from both the high prices obtained for their fruit and a more favourable rate of exchange.

Citrus experts are of the opinion that citriculture should be restored to its former proportion, by the rehabilitation of existing groves, an increase in yields to 140 cases to the dunam already obtained in high-quality orchards—or at least 100–110 cases to the dunam instead of the present average of 70 cases—and finally by the planting of new groves, on an area of 80,000–100,000 dunams.

Citriculture in Israel enjoys the advantages stemming from highly efficient organization, owing to the existence of an authoritative controlling body—the Citrus Marketing Board—and the excellent arrangements made for the export of the fruit. By virtue of the protection extended to the Board by law, and the prestige it enjoys as a result of the Minister of Agriculture serving as its Chairman, the Board wields a considerable measure of influence over the rehabilitation of groves, methods of cultivation, the institution of centralized packing and planned marketing, both within this country and abroad.

LIVESTOCK

Jewish settlers in this country, when they renewed their ties with the soil after an interval of seventy-five generations, were compelled not only to accustom themselves to manual labour and to adapt themselves to cultivation of the soil, but to restore the natural fertility of the waste lands, and to make them more responsive to their efforts.

They rejected the primitive methods of agriculture practised at that time in Palestine and proved willing and apt pupils of all who could teach them the art of modern farming. They studied rational methods of cultivation and a proper rotation of crops, the use of suitable and efficient implements, the eradication of pests and weeds, the widest possible use of irrigation, the introduction of new cultures and varieties and better methods of co-operation, organization and research. Innumerable difficulties beset their path and present achievements represent the result of three generations of unremitting effort.

The native breed of cattle which Jewish pioneers found upon their arrival in this country, barely existing on the scanty pastures, could not produce more than several hundreds of litres of milk annually. The new settlers sought to acclimatize a more productive breed.

The wool of the local breed of sheep was coarse, its meat of low quality and the yield of milk poor and suitable only for the production of cheese. For an entire generation, Jewish settlers endeavoured to improve this breed, and today their flocks average hundreds instead of the previous dozens of litres of milk every year.

The case of the native breed of poultry was similar. The scrawny local fowls, scratching in the manure pile or in the rubbish heap, never laid more than a few dozen tiny eggs every year. The new strains introduced produced three times as many in number and no less than five times in weight.

Coastal net fishing was not attractive to the young generation, to whom the deeper waters beckoned. New equipment was acquired—boats and trawlers, drag-nets and lights and echo-sounder equipment. After the first generation of Jewish fishermen had been initiated into the mysteries of the sea, their successors began to make a closer study of the fisherman's craft in schools specially established for that purpose. Moreover they learnt that the carp, much favoured on the Jewish table, could be bred on land, under conditions of artificial feeding, in special ponds.

But the livestock branches of Israeli farming are now experiencing a crisis—not a biological, nor a zoo-technical, nor even an economic crisis—but a crisis of planning. The present proportions of the cattle herds are being subjected to criticism. It has been suggested that further expansion be restricted, not because milk is not popular with the Jewish consumer, but because the country's economic straits make the dairy too expensive a source of animal proteins, and because the cow requires a large area of land. It would be far better, members of this school hold, to devote the limited area at our disposal to the production of vegetable products, for direct human consumption from which cheaper proteins could be extracted. We must produce the calories we need by fostering crops rich in starch and sugar, like the Japanese whose staple food is rice, rich in calories, and not animal proteins.

A similar complaint is directed against poultry, which are voracious consumers of grain. Here, too, it is claimed, it would be advisable to raise these grains for direct human consumption, without the aid of the chicken to convert them into protein-rich eggs.

No criticism of the sheep has been made except that it multiplies so slowly—but this is not the fault of the planners.

The development of marine fishing has encountered serious obstacles. Fishing is an occupation calling for a high degree of devotion, a willingness to accept the hardships of life on the sea, away, for lengthy periods, from hearth and home. Also it requires very large investments, but above all the necessary knowledge of the fishing grounds is still lacking.

The advisability of developing the breeding of fish, too, is being queried, for the ponds require a large volume of sweet water which might be more profitably used for irrigation.

But despite this criticism all of these branches of livestock farming have persisted and even progressed. It was precisely in these five years, when the development of agriculture was planned to a greater extent than previously, that local herds of cattle doubled, despite the fact that they had to be supplemented by imports from abroad, and notwithstanding the seductions of the black market for beef. In this period local flocks of sheep also doubled.

Marine and coastal fishing, lake fishing and carp-breeding, have all expanded and increased two-fold over the five years under review.

Even apiculture, neglected because of its demands for a relatively small quantity of sugar, has produced more honey in the past two years than in the previous period.

Only in chicken farming has growth been arrested, though it, too, has held its ground despite the restrictions to which it has been subject, the various controls imposed, the decreased import of feeding stuffs and the lack of credit.

Cattle

East-Friesian cattle were introduced into this country by Jewish settlers in the year 1922, and by crossing with the Damascus breed, the foundations of the high-quality herds which distinguish Israel's dairies today were laid. The import of East-Friesian bulls has continued throughout these years in order to maintain the high standards achieved. Pedigree cows are also imported from time to time. Intensive methods of dairy-farming in this country have resulted in a higher milk yield from the cross-bred cow than that of this breed in other countries. After twenty-five years of careful breeding local herds totalled 36,043 head in 1947, and six years later—69,781 head. The rate of increase is reflected in the following table:

TABLE 59

Cattle (no. of head)

| | December 1947 | Cattle yearly average[1] | | | | | | December 1953 |
		1948	1949	1950	1951	1952	1953	
Total no. of cattle	36,043	33,890	35,759	44,599	55,302	61,910	67,326	69,781
Cows	21,105	19,559	19,978	24,390	29,005	33,621	36,004	37,236
Heifers	4,875	4,559	5,145	6,064	8,559	8,396	8,016	8,050
Calves	9,708	9,440	10,247	13,582	17,013	19,149	22,448	23,711
Bulls	316	272	289	364	414	446	437	443
Bull calves	39	60	100	199	311	298	421	341

From the above table it will be seen that during the War of Liberation the number of head of cattle declined. The actual decrease was even greater. In December 1948, the total was 33,576, or 2,467 head less than at the end of the previous year, as a result of losses caused by enemy shelling and other military operations. The energetic efforts to rehabilitate the herds is also indicated and only a year later the number of head totalled 37,523, or 1,480 head more than in December 1947. The growth of the herds at an accelerated pace was mainly due to imports from Holland and the United States. In 1950 the increase was 9,000 head and in the following year 11,000 head; thus over a period of five years the number of cattle doubled. The significance of this rate of increase is underlined by the fact that in the previous five years, from 1942 to 1947—also a period of agricultural expansion—the herds increased by only 25%.

The increase in yield is given in the following table:[2]

TABLE 60

Milk production (kl.)

	1948/49	1949/50	1950/51	1951/52	1952/53	1953/54
Gross yield	80,150	96,600	108,500	125,950	139,200	163,000
Cow's milk	72,200	87,500	97,500	113,200	121,800	140,000
Sheep's milk	7,950	9,100	11,000	12,750	17,400	23,000

In spite of this steady increase domestic supplies of milk were not sufficient to meet consumption and the Government was compelled to eke out the volume of pure milk available by the addition of milk powder (mainly skimmed milk powder) and the distribution of hard

[1] *Statistical Monthly for Israel*, Part B, Vol. II, February 1954, p. 96.
[2] Files of Cattle Breeding Division, Ministry of Agriculture.

cheese and imported butter. Thus during the seasonal decline in yield milk was adulterated upon Government instructions and the fat content reduced from 3·5% to 2·3%. Producers complain that this practice has reduced the demand for milk, pointing to the 1947 figures when the average per capita consumption was 135 litres as compared with 77 litres in 1953. According to an official estimate[1] the per capita consumption of milk and milk products was 96·3 litres in 1949/50, 96·7 litres in 1950/51 and 93·6 litres in 1951/52. While producers complain that this decline has been caused by the low quality of adulterated milk, it is stated that the real reason has been

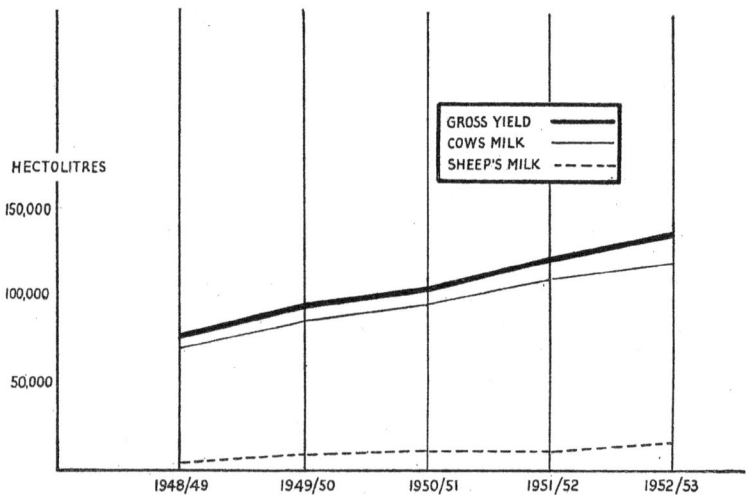

Fig. 4.—Yield of Milk (1948/49–1952/53)

the reduced purchasing power of the population. It seems that there is an element of truth in both statements. But in view of the acute shortage of foodstuffs rich in animal proteins such as meat, fish and eggs, the consumption of milk, which is highly nutritive, especially for the lower age-groups, must be encouraged. If per capita consumption were restored to its 1947 level of 135 litres annually and the practice of adulterating milk was discontinued, a quantity of 270 million litres would be required for a population of two millions. Taking into account an increase of 5% required for rearing, Israel's herds of milch cows would have to be increased to 72,000 head, or double the present number. In addition we must reckon with an

[1] *Statistical Monthly for Israel*, Vol. IV, No. 6, November 1953, p. 862.

increase in the number of heifers, calves, bulls and bull calves, according to the usual proportion of 80% of the number of cows. Thus local herds would have to number 130,000 head. Even then the average of 65 head of cattle to every thousand of population would be among the lowest in the world, but it would be higher by 44% than the present figure. Dairying can and must make a far greater contribution towards balancing the national diet, but this implies maintaining the present composition of field crops, concentrating upon the production of fodder, even if it should reduce the prospect of increasing the production of bread-grains. It seems to us that should such a choice have to be made, milk production should be given priority.

Sheep

The fact that flocks of sheep on Jewish farms have increased four-fold over the past five years, does not give a true picture of the achievements of sheep-breeding, nor of the aims and the planning of this branch. Prior to the foundation of the State of Israel, Arab flocks comprised 200,000 head of sheep and 300,000 goats. The yield of milk per head of sheep and goat was low. The goat co-operated with its owners in destroying the soil of the country, and in leaving it unprotected to the ravages of erosion by rain and wind. In their flight from this country the Arabs took their flocks with them. The virtual disappearance of goats, which had previously roamed and grazed at will, was soon noticeable when areas that had been denuded of vegetation were suddenly covered with sparse verdure and stunted bushes, the shoots of which had previously been devoured regularly in the spring by the goats.

The fact, however, that quite recently the country supported hundreds of thousands of sheep and goats, constituted the basis for the planning of this branch, though the line of development chosen was peculiar to Jewish breeders.

In January 1954 a census of flocks in Israel was carried out. The census was complete with the exception of flocks owned by immigrants from Kurdistan, who refused to submit the necessary figures.

The collective settlements were the pioneers of sheep-breeding in the Jewish sector of agriculture, and they made persistent efforts to introduce it into private farms too. They were not very successful in this respect, but as a result of radical changes in local farming, following the establishment of the State of Israel, an improvement was registered and private breeders are now making a substantial contribution towards the development of sheep-breeding.

The following table indicates the distribution of flocks (January 1954):[1]

TABLE 61

Sheep (flocks and head)

	Flocks	Head	%
Private breeders	174	13,658	22·3
Collective settlements	113	37,126	60·7
Co-operative and other villages	33	8,682	14·2
Farm schools	10	1,456	2·4
Experimental stations	2	248	0·4
Total	332	61,170	100·0

Breeders in this country call a unit comprising at least 50 head a flock. A smaller unit is called a group. The following was the distribution of the sheep during the period of the census.

TABLE 62

Distribution of sheep

Less than 50 head	58 groups	17%
51–100 ,,	72 flocks	22%
101–200 ,,	67 ,,	21%
201–300 ,,	50 ,,	15%
301–400 ,,	57 ,,	17%
401–500 ,,	22 ,,	6%
More than 501 head	6 ,,	2%
Total	332 flocks	100%

Smaller flocks are now beginning to be a common feature of Jewish settlements, and it is gratifying to note that some private breeders have larger flocks of sheep.

The veteran breeders are making constant efforts to improve the quality of their flocks, and a flock book, run on professional standards, is being kept. It is these breeders, mainly, who are responsible for raising the standards of sheep-breeding in this country. Their example is being followed by others who keep card indexes, select their dams carefully, pay special attention to problems of breeding, submit to periodical milk tests, etc. These breeders own more than half of the flocks on Jewish farms. There is also a third class, whose flocks are being built up in keeping with the instructions of the Ministry of Agriculture's Sheep-breeding Division, and the Sheep-breeders' Association, but do not yet submit to milk tests.

[1] Data concerning the Census of Sheep and the Pedigree Book were supplied by courtesy of the Director of Sheep-breeding Division, Ministry of Agriculture.

The distribution of the country's flocks according to the above classification is reflected in the following table:

TABLE 63

Classes of flocks

Class A	33 flocks comprising		12,281	head
Class B	120	,,	,,	31,123 ,,
Class C	179	,,	,,	17,766 ,,
Total	332	,,	,,	61,170 ,,

Flocks of the first and second classes are entitled to belong to the 'Flock Book' or 'Pedigree Book' established for local sheep. Of the 153 flocks in these two categories, 89 flocks were so registered at the time of the census.

The development of sheep-breeding as reflected in the figures of the Flock Book is shown in the accompanying table:

TABLE 64

Flock Book in 1951/52–1952/53

Year	No. of flocks registered in Flock Book	No. of Ewes	Average yield (kg.)	Record yield (kg.)	Ewes giving more than 250 kg. No.	%	Yield above 400 kg. No. of ewes
1951/52	75	15,178	239	650	6,708	44	534
1952/53	88	18,522	240	650	8,302	45	648

It must be borne in mind that these results were obtained with the local Awassi breed, which under primitive conditions produces no more than 50 kilograms of milk yearly. The yields indicated above —averaging five times more than previous yields of the local breed— have been achieved chiefly as a result of careful breeding. Of course these efforts to improve the stock do not end with raising of the annual yields. They are being continued, the short-run objective being to raise it above the 300-kilogram mark.

In the table that follows we can trace efforts to improve local sheep over a period of seventeen years, for the eleven best flocks in the country. We can compare them with the 1937/38 basic year.

The average annual milk yields of these eleven flocks, counting 2,881 ewes, exceeded 300 kilograms. The number of ewes whose yield exceeded 250 kilograms rose from 0·62% in 1937/38 to 80% in 1952/53, while the average yield rose in this same period by 140%. This is indeed a fitting reward for progressive breeding.

The flock of one veteran collective settlement, which pioneered efforts to improve the local breed, included—in 1951/52—497 ewes

TABLE 65

Progress in milk yields in the period between 1937/38 and 1952/53

Year	No. of ewes in 11 flocks	200 kg. Ewes	200 kg. %	200–250 kg. Ewes	200–250 kg. %	250–300 kg. Ewes	250–300 kg. %	300–350 kg. Ewes	300–350 kg. %	350–400 kg. Ewes	350–400 kg. %	Above 400 kg. Ewes	Above 400 kg. %	Above 250 kg. Ewes	Above 250 kg. %	Record yield (kg.)	Average yield (kg.)	Increase compared to 1937/38 kg.	Increase compared to 1937/38 %
1937/38	975	892	91·49	77	7·89	6	0·62	—	—	—	—	—	—	6	0·62	290	130·95	—	—
1942/43	1,944	1,061	54·54	565	29·06	240	12·34	67	3·45	10	0·51	2	0·10	319	16·40	430	188·10	57·15	43·62
1943/44	2,005	822	41·00	614	30·62	372	18·55	162	8·08	30	1·50	5	0·25	569	28·38	490	210·22	79·27	60·51
1944/45	1,962	546	27·83	583	29·71	464	23·65	257	13·10	85	4·33	27	1·38	833	42·46	490	232·74	101·79	77·72
1945/46	2,117	468	22·11	578	27·30	557	26·31	367	17·33	117	5·53	30	1·42	1,071	50·59	480	245·65	114·70	87·56
1946/47	1,779	227	12·76	421	23·66	472	26·54	370	20·80	185	10·40	104	5·84	1,131	63·58	610	274·31	143·36	109·48
1947/48	1,941	270	13·91	361	18·60	482	24·83	395	20·35	235	12·11	198	10·20	1,310	67·49	650	284·22	153·27	117·04
1948/49	2,255	216	9·58	347	15·39	537	23·81	472	20·93	333	14·77	350	15·52	1,692	75·03	680	304·65	173·70	132·65
1949/50	2,268	183	8·07	335	14·77	602	26·54	488	21·52	318	14·02	342	15·08	1,750	77·16	680	305·25	174·30	133·10
1950/51	2,442	176	7·21	377	15·44	558	22·85	562	23·01	369	15·11	400	16·38	1,889	77·35	650	308·94	177·99	135·92
1951/52	2,709	243	8·97	346	12·77	647	23·88	649	23·96	418	15·43	406	14·99	2,120	78·26	650	306·84	175·89	134·32
1952/53	2,881	190	6·59	385	13·36	619	21·49	734	25·48	493	17·11	460	15·97	2,306	80·05	650	314·14	183·19	139·89

averaging over 338 kilograms annually, and achieving a record of 610 kilograms; 27% of the ewes in this flock produced over 400 kilograms annually. This settlement increased its average yield by 225% over a period of seventeen years, establishing a new record average yield.

Flocks in Israel are small. There are no breeders in this country on the scale common in New Zealand or Australia. Here we have a rare case of sheep-breeding for milk production, while in other countries sheep are kept for either wool or mutton. The keeping of sheep for milk is more complex, the flock requiring greater care in respect of feeding, housing and milking. Wool and mutton in this country are by-products of sheep-breeding.

Poultry

A major contribution was also made by Jewish settlers in the field of poultry-raising. In the initial period of Jewish settlement, the pioneers kept the small, multicoloured local chicken, whose yield was low and reproduced by direct hatching. The early settlers, like the Arab peasants, despised poultry-raising. It was only in 1922 that various other breeds were introduced from abroad and chickens distributed among Jewish farmers. Experiments were launched in methods of feeding, research was conducted in poultry diseases and a systematic disease control was established. A staff of experts was trained and the work of improving strains and cross-breeding between the local fowl and White Leghorns was begun. Later the runs were stocked exclusively with the latter. Special hatching farms, conducted under suitable controls and supervision, were established, standards for laying were set, a Pedigree Book was inaugurated and a Union of Hatching Farms, affiliated to the Poultry-breeders' Association was formed. In 1938 the Union began to publish the results of its work. In that year every fowl, placed at the beginning of the year under control, averaged 96·30 eggs. Despite the difficulties encountered during the war years this figure was raised to a yearly 140 eggs fourteen years later.

The Hebrew University and the Veterinary Division of the Government of Israel have carried on unremitting research work with a view to combating the diverse poultry diseases and reducing mortality in the runs. As a result of these efforts mortality has been reduced by 15%.

In 1927, when the first census of Jewish agriculture was conducted, 25,681 chickens were registered. Twenty-one years later, when the State of Israel was established, the number of laying hens was 1,400,000, the figure being almost doubled within a year. Production rose from three million eggs to 400 million, in addition to 7,500 tons of dressed and live poultry. It was during this period that poultry-breeding ran into economic difficulties that made its further expansion at least problematical. These difficulties have little to do with problems of marketing. In Israel eggs and poultry are highly favoured by the Jewish consumer. To supply a daily minimum of one egg per person five million birds would be required, and even at this level it is doubtful if the market would be satisfied. The difficulty stems from the fact that a million birds consume 55,000 tons of feed annually, including expensive imported feeding stuffs, such as grains, fish-meal, bran, oil-cake, minerals and vitamins. The expenditure of large sums in foreign currency on the import of these feeding stuffs is neither economically justified, nor within the financial capacity of the State.

Some years ago the Government of Israel introduced a system of rationing feed and the expansion of poultry-breeding has been checked as a result.

At the same time it is agreed that poultry-breeding must be maintained and even allowed to develop to meet the demands of the market, provided one stringent condition is complied with—that it be based upon locally produced feed. Is this possible?

In an earlier chapter, in our discussion of field crops the difficulties attending the cultivation of food crops, particularly for bread, were emphasized. Cattle, sheep and draught animals are all major consumers of vegetable products. At the same time the choice of alternatives is narrowly restricted by the limited area of land at our disposal. Is there any way out of this quandary?

A solution is possible provided the farms can produce a substitute for feeding grains requiring a minimum area for cultivation. Preliminary tests indicate that the carob bean can solve the problem both in so far as nutritive value and prospects of unrestricted cultivation to meet the needs of poultry-farming are concerned. The carob is a native of very long standing in this country. It bears fruit two or three years after grafting and reaches full bearing within five to eight years. A yield of one ton of carob beans from every four trees can be achieved without any difficulty. Every breeder of fifty chickens can easily grow four trees and thereby obviate the necessity to purchase feeding grains. One million trees can provide feed for six million chickens within a comparatively short period. These plantations will not encroach upon any cultivable land. Neither time nor energy should be wasted trying to persuade breeders to grow carob trees. They have not done so up to the present because they cannot be expected to wait years for a crop. Nor should carobs be allowed to occupy any of their valuable farm land, which is required for other crops. Carob plantations already exist. Hundreds of thousands of trees, perhaps even the entire million required, are scattered all over the country in the highland regions and along the coast. All that is required is some central authority to undertake cultivation of the trees, the planting of new plantations, the picking and the marketing of the fruit. Just as there already exists a central authority for afforestation—or rather two, one controlled by the Government, the other by the Jewish National Fund—one or both of these, or perhaps a third authority, such as the National Orchards Company, should undertake the organization of carob cultivation, thereby providing the solution to the problem which is jeopardizing the future of poultry-farming.

The expansion of ground-nut growing for the production of oil will help supply oil-cake for the chickens which is another problem. Lucerne meal, supplemented by sweet potato meal (which must be produced), sugar-beet waste from the sugar factories, bran from the flour-mills, minerals already available in this country, the development of fisheries, can all, given efficient organization, contribute towards the consolidation of the agricultural and economic basis of poultry-farming in Israel.

Apiculture

One of the early settlers, who seventy years ago, in the year 1884, placed a collapsible hive in his farmyard at Ness Ziona, was the pioneer of modern bee culture in this country. Previously honey had been elaborated from earthern jars. Before the turn of the century the example of Ness Ziona had been followed by all the settlements of the Lowland, which had installed modern hives.

In 1920 there were 1,500 hives in Jewish settlements; by 1930 this figure had risen to 4,000, by 1940 to 18,000, by 1950 to 29,000 and totalled 35,000 in 1953.[1]

Apiculture in Israel has always had to contend with rot and the depredations of wasps. Beekeepers suffered considerably as a result of operations during the War of Liberation when, because of the hostilities, the fallowing of large areas of land and an exceptionally severe attack of wasps, thousands of hives were depopulated. Despite this setback, however, the keeping of bees has progressed, albeit gradually.

The number of beekeepers is 700, of whom 580 are individuals, 80 are collective settlements and 40 are schools, farming estates, experimental stations, etc.

About two-thirds of the hives are concentrated in the citrus belt, the remaining third being scattered over other parts of the country.

[1] Lists of the Beekeeping Section, Ministry of Agriculture.

The number of hives kept varies with the type of keeper. In the collective settlements the number is 150–200, private farmers keep on an average 50–60, while amateurs rest content with 10–20. There are also a number of expert keepers who own between 600–800 hives. The yield of honey over the five years being reviewed was as follows (in tons): 500, 420, 400, 580, 700, respectively. The yield fluctuates with seasonal conditions, the blossoming, diseases, etc.

Jewish beekeepers have not rested content with the modernization of their technical equipment, but have penetrated into the biological sphere and have discovered that the local variety suffers from an angry disposition, making it difficult for newcomers to take up bee-keeping. They have also a strong tendency towards swarming, reducing the yield. Efforts to secure a more suitable strain have been comparatively successful, following the introduction of high-quality Italian queens. The breeding-stock has been widely distributed and already covers a third of the keepers in this country. The results are already apparent, for the Italian breed is quieter, and its yield is higher, by 50%–100%, than that of the local breed.

The immediate purpose of beekeeping is, of course, the production of honey. But the importance of the bee is mainly in the fertilization of vegetation resulting in increased yields. For this reason beekeeping must be extended over all parts of the country.

Meat-production

Meat-production, as a branch of agriculture, conducted extensively, upon a scale common in countries with rich natural pastures, is not possible in Israel. In Israel the production of meat is ancillary to dairying, sheep-breeding and poultry-raising. The poultry runs are Israel's most important suppliers of meat, marketing 73% of the total local production, over the past six years.

The following table presents the contribution of each branch to the gross production of meat.

TABLE 66

Production of meat (tons liveweight)[1]

	1948/49	1949/50	1950/51	1951/52	1952/53	1953/54
All meat	7,150	9,250	9,400	9,500	11,050	13,650
Beef ⎱	2,150	1,950	2,300	2,550	2,500	3,600
Mutton ⎰					550	850
Poultry	5,000	7,300	7,100	6,950	8,000	9,200
Poultry (%)	70	78	75	73	72	67

[1] *Statistical Abstract of Israel*, 1953/54, p. 74.

The quantity of beef marketed has increased in recent years mainly as a result of the development of a new branch of farming—the fattening of cattle—the progress of which is reflected in the following figures for the years 1950–54 (spring) in the number of heads kept for this purpose: 316, 267, 574, 1,795, 2,897.

The calves for fattening—previously slaughtered at an age of several weeks—are kept for 18–24 months until they reach a live-weight of 250–400 kilograms.

Hereford and Brahma bulls and cows are being imported to lay the

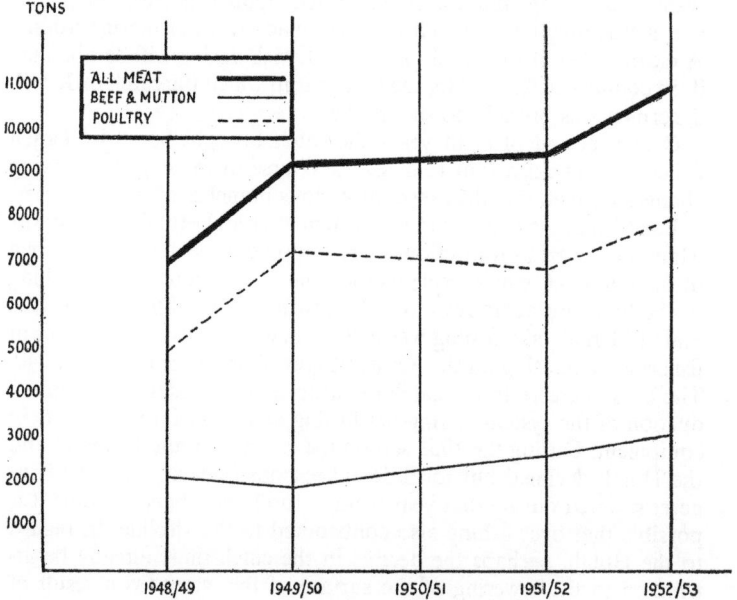

Fig. 5.—Production of Meat (1948/49–1952/53)

foundation of local beef-production. This initiative is highly important, though of necessity, in view of the limited area available for grazing, it will be restricted. The raising of beef-cattle on a larger scale must, of course, be conducted extensively to produce meat at reasonable prices.

FISHERIES AND FISH-BREEDING

Current per capita consumption of fish in Israel averages 13–14 kilograms annually, having declined from the previous per capita figure of 16 kilograms. Of the 21,000, 22,000 and 27,000 tons of fish respectively consumed in this country during each of the three years 1951–53, the percentage caught or bred locally was 34%–35%. To supply the population of Israel with an annual average of 16 kilograms per capita, it will be necessary to increase its fisheries by 300%.

There are three distinct branches of Israeli fishing—marine fishing, lake fishing and fish-breeding. Marine fishing, again, is further divided into three sections, inshore, pelagic and deep-sea fishing.

In the initial period, when Jews in this country first began to penetrate into fishing, interest concentrated upon fish-breeding ponds, and it was only later that efforts were made to enter marine fishing. A clearer idea of the development of Jewish fishing will be obtained if we commence from 1945, the first year in which the catch of Jewish fishermen was brought to the market.

Over a period of eight years the catch brought home by Israeli fishermen has risen four-fold. In the course of these years, striking changes have altered the aspect of various branches of local fisheries.

Lake fishing is conducted in the Kinneret and Hulah. During the Mandatory period restrictions were imposed upon Jewish fishermen in the Kinneret, which explains the small dimensions of lake fishing in the first four years reviewed. Following the establishment of the State of Israel, lake fishing was given a new stimulus. The number of fishermen operating on the Kinneret grew from some dozens to 240. The large increase in the catch of sardines is the result of the introduction of the system of ring-net fishing and other improvements in equipment. During the 1953 season the catch from the Kinneret and the Hulah declined but for different reasons in each case. The Kinneret suffered during that year from natural disturbances, while it is possible that over-fishing also contributed to this decline. In regard to the Hulah, perhaps the decline in the catch must already be attributed to the lowering of the surface of the water (as a result of extensive drainage works), as well as excessive use of the electric method of fishing. Here too the possible influence of climatic disturbances must be taken into account. However, on the basis of one year's experience it is too early to draw any final conclusions.

Under the Mandate, coastal fishing was confined to the Haifa Bay, and was virtually an Arab monopoly, Jews being to all intents and purposes excluded. Following the establishment of the State ex-servicemen and new immigrants formed fishing groups, which were based upon Achzib, Naharia, Acre, Tirah, Athlit, Gesher Zarka, Tel-Aviv/Jaffa and Nebi Rubin. The return of Arab fishermen to Acre, Gesher Zarka and Haifa came at a later stage. The latter are organized in co-operatives while there are a number of mixed Jewish-Arab fishing groups.

TABLE 67

Israel fisheries 1945–53[1]

Year	Total		Lakes			Inshore and pelagic fishing			Deep-sea fishing			Fish-ponds		
	Tons	Index	Tons	Index	%	Tons	Index	%	Tons	Index	%	Tons	Index	%
1945	1,836·5	100	140·2	100	7·6	9·9	100	0·5	439·0	100	23·9	1,247·4	100	67·9
1946	1,759·7	96	120·1	92	6·8	7·5	76	0·4	307·2	70	17·5	1,324·9	106	75·3
1947	2,061·9	112	102·9	73	5·0	1·8	18	0·1	227·4	52	11·0	1,729·8	139	83·9
1948	2,703·0	147	141·0	100	5·2	16·4	166	0·6	110·9	25	4·1	2,434·7	195	90·1
1949	4,233·6	231	396·7	283	9·4	252·4	2,550	5·9	646·3	147	15·3	2,938·2	236	69·4
1950*	7,087·1	386	707·0	504	10·0	788·2	7,962	11·1	1,091·9	249	15·4	4,013·5	322	56·6
1951†	8,228·0	448	927·2	661	11·3	1,071·4	10,822	13·0	928·6	212	11·3	3,903·5	313	47·4
1952	7,493·4	408	1,046·0	746	14·0	1,045·1	10,556	13·9	953·2	217	12·7	4,449·1	357	59·4
1953	7,657·5	417	781·3	557	10·2	938·2	9,476	12·3	1,286·1	293	16·8	4,651·9	373	60·7

Fishing in distant waters, 486·5 tons (6·9%) must be added. † Fishing in distant waters, 1,397·3 tons (17·0%) must be added.

[1] *Israeli Fisheries*. A monthly for information and statistics, Ministry of Agriculture, Tel-Aviv, 1954, p. 37.

Fishing in upper waters was commenced only in 1940. In that year the catch totalled several hundreds of tons as compared with the insignificant quantity which had been caught previously. Progress in this type of fishing may be attributed to three different causes. Groups of experienced Jewish fishermen came to this country from Tripoli, bringing with them their equipment. They introduced the improved light methods of fishing. Their example was followed

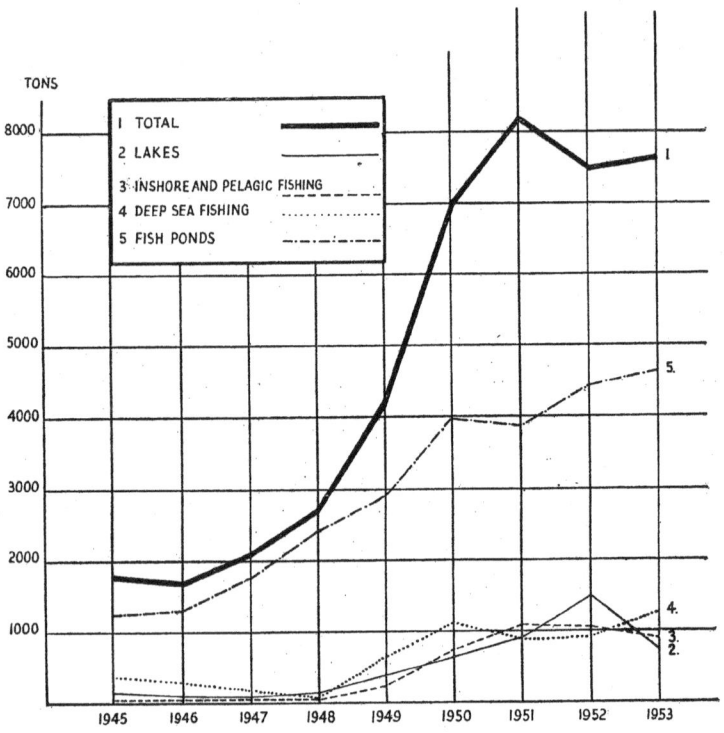

Fig. 6.—Sources of Israel Fisheries (1945–53)

by other fishermen, who used the Italian method of light fishing. The crown-net method was introduced by two groups from Turkey. While this method can be further improved by mechanization, its effectiveness was immediately apparent.

In 1950 the Government of Israel purchased three Italian trawlers, fitted with ring-nets. Half of the crews were Israeli fishermen. The trawlers were joined by two Israeli vessels who used the same system.

The rise in the catch of upper-waters fishing must be ascribed mainly to the introduction of the ring-net method, light and crown-net fishing methods being contributory factors.

The slight decline registered last year is attributed to the crisis which is confronting sardine fishing throughout the world and compelled ten groups, who have been fishing in upper waters as well as engaging in trawler fishing, to discontinue operations. Israeli fishing is mainly based upon its fleet of trawlers, the expansion of which though gradual has been maintained throughout the five years under review. In 1948 four trawlers were in operation but by the close of that year the number had risen to fourteen. In 1949 twenty-eight trawlers based upon Israel were engaged in fishing. Six of these were Italian boats, while two of the Israeli boats were wrecked. In 1950 six new trawlers were purchased but two were laid up permanently. In the following year all the Italian vessels left the country. Two new vessels were purchased and five small, leaky boats were laid up. In 1953 another two vessels were taken out of commission, leaving the trawler fleet with nineteen vessels. Meantime orders for new trawlers have been placed abroad and according to plan eighteen or nineteen new vessels will join the fishing fleet round the end of 1954 and the beginning of 1955 bringing the total number of trawlers up to thirty-seven or eight.

Trawler fishing can consolidate its achievements and expand provided the requisite effort is made and funds are invested in renewing the fleet, improving the equipment of the fishermen, providing them with better living conditions at sea and by raising professional standards.

Fish-breeding

Towards the close of 1936 the first Jewish Settlement was established in the Beth Shean Valley. Within the next two years it was followed by eight others, the prospects of all being doubtful. Water was abundant but saline, the soil had suffered from long neglect, while agricultural conditions were unfavourable. It was clear that supplementary sources of income must be developed. In one of these villages, a young settler had embarked upon the breeding of carp in artificial ponds. Spawn was imported from abroad and was introduced into artificial ponds, filled with water from the many springs of the district. The fish were fed on lupins, and by 1940 there were 150 dunams of ponds in four settlements, the average annual catch being 93 kilograms to the dunam. Within the short space of five years the area under ponds increased to 7,950 dunams, with an annual catch totalling 1,153 tons, averaging 145 kilograms to the dunam. By 1950 the area had increased almost three-fold to more than 21,000 dunams, the catch totalling 3,700 tons, an average of 176 kilograms to the dunam. Since then there has been a steady advance in respect of both the area under ponds and the total catch, though the average per dunam has declined owing to a shortage of feeding stuffs which must be imported from abroad.

Criticism has been voiced regarding the expansion of fish-breeding and the location of the ponds, because of the excessive volume of water required (more than ten times as much as summer crops), because of the fact that not infrequently good land that could be more advantageously used for other purposes is utilized for the construction of ponds, and finally because of the dependence of fish-breeding upon imported feeding stuffs which must be paid for in foreign currency.

Obviously if fish-breeding were concentrated in areas with abundant water resources and used only sub-marginal soil it would merit general encouragement. The regulations governing the regional distribution of the ponds are not adequately observed and in each successive year fish-breeding encroaches upon new areas.

TABLE 68

Regional distribution of fish-breeding ponds (October 1953)[1]

	Area	
District	Dunams	%
Upper Galilee	10,987	33·1
Jordan Valley	3,304	10·0
Beth Shean Valley	7,887	23·8
Jezreel Valley	1,493	4·5
Zebulon Valley	4,702	14·2
Coastal Plain	4,784	14·4
Total	33,157	100·0

The Seven Year Agricultural Development Plan sets the limit of fish-breeding ponds at 40,000 dunams. This goal is justified but it is highly desirable that expansion should be restricted mainly to the Beth Shean Valley and not be permitted in other districts where fish-ponds use sweet water and good agricultural land which can be more profitably utilized for other purposes. The 200 million cubic metres of water these 40,000 dunams of ponds will require annually would suffice to irrigate an area of 300,000 dunams under agricultural crops.

Tests are at present being conducted to establish whether the huge Beth Netupha Reservoir can be stocked with fish. If these experi-

[1] *Israeli Fisheries in 1952/53.* Statistical Report, Ministry of Agriculture, Tel-Aviv, 1954, p. 20.

ments should prove successful it will be possible to embark upon large-scale breeding of fish. The water stored in the Reservoir is of course intended for irrigation purposes and thus no wastage will be involved.

We have reviewed the main branches of farming in Israel, tracing their growth and development over the past five years, noting the contribution made by the new immigrants in settling the land, in extending the bounds of Jewish agriculture and in increasing farm production.

But the progress that has been made must be appreciated in terms of national construction. Obviously much time is taken up in the laying of foundations and the building of the substructure. Once this

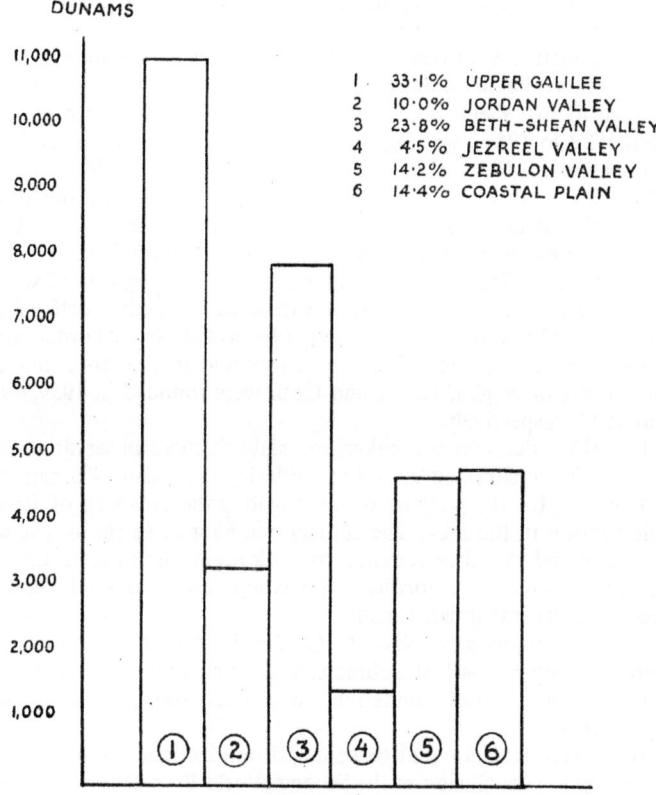

DUNAMS

1	33·1%	UPPER GALILEE
2	10·0%	JORDAN VALLEY
3	23·8%	BETH-SHEAN VALLEY
4	4·5%	JEZREEL VALLEY
5	14·2%	ZEBULON VALLEY
6	14·4%	COASTAL PLAIN

Fig. 7.—Regional Distribution of Fish-Breeding Ponds (October 1953)

task is completed the rate of advance may be accelerated. A niche in Israeli history is reserved for the pioneers of Jewish agricultural settlement, for the innovators, the first husbandmen and herdsmen who laid the foundations of modern farming in this country. Ultimately the present rapid rate of progress must be credited to their pioneering efforts.

One of the most spectacular and daring chapters of this tale of colonization, from the political no less than from the agricultural aspect, was the attempt to evolve suitable methods of settling the arid wilderness of the Negev.

The Negev is the dry-land, the land of drought. At its southern tip, in Elath, the quantity of rainfall is insignificant, hardly more than some millimetres throughout the year. In 1952 no more than 17.3 millimetres were registered. In the Red Sea region, indeed, there is no data available on annual rainfall exceeding 50 millimetres. The average in Beersheba taken over a period of twenty-three years—from 1922 to 1945—was 202 millimetres, with a minimum of 129·8 millimetres in 1928/29 and a maximum of 336·1 millimetres in 1933/34. A belt, 15 kilometres wide, north of Beersheba, constitutes the border of the arid zone. Normally, however, the borders of the Negev are extended to include an area where annual rainfall fluctuates between 250 and 350 millimetres and the prospects of winter crops are unstable. But this area, extending along the north of the Negev, should more properly be regarded as the Negev border zone. It was with the object of gaining a foothold in this area that the settlements of Negba, Dorot and Gath were founded in 1939, 1941 and 1942, respectively.

In 1943 a decision was taken to establish three observation stations, to be constructed upon lines differing from that of permanent settlements, for the purpose of investigating the prospects of Jewish colonization in the area. The settlers manning these stations it was contemplated would be required to work according to plans drafted by the settlement authorities, who would take the final decision regarding the investments made.

The three stations established—Gvulot, Revivim and Beth Eshel—were to study the soil, the climate, vegetation and water resources, with a view to the foundation, at a later stage, of permanent settlements.

But before results could be achieved political conditions changed following the conclusion of the Second World War. It became necessary to establish new settlements to counter the Land Laws, under the provisions of which the Negev was a prohibited zone for Jewish settlement. Thus in October 1946, in a single operation, seven new settlements were founded in various parts of the Negev. From the purely economic point of view settlement work in this drought-stricken zone was not justified, at least until some solution to its greatest problem, the lack of water, had been discovered. The Negev covers an area of more than 12 million dunams or about 60% of the territory of the State of Israel. There was no reason to assume that even if the quest for subterranean water resources proved successful, the volume of water would be large. The alternative was to bring water from a distance. This solution, too, could hardly be adequate to the problem presented by the waterless Negev. According to hydrological data at our disposal, Israel's water resources cannot be expected to exceed three billion cubic metres—sufficient, at the very most, for the irrigation of three million dunams. Immense irrigation works including canals running down the entire length of Israel from north to south will be required to create a handful of oases in this wilderness.

The execution of such a plan in its entirety involves an immense investment and will require between fifteen and twenty years before it can be completed. The various stages of the overall plan have therefore been determined according to the sources of water that can be tapped. The primary source is subterranean water, closer to the Negev, in the vicinity of Nir Am and Gvar Am and further north along the coast, bordering the Migdal–Faluja road. Assuming that it is possible to pump four million cubic metres from every kilometre of the coast, it has been estimated that 45 million cubic metres can be obtained from this source.

The second source is the Yarkon River, whose flow, totalling 250 million cubic metres annually, empties mainly into the Mediterranean. The plan provides for the supply of a certain volume of water to the immediate neighbourhood of the river, Tel-Aviv and Jerusalem, leaving a total volume of 125–150 million cubic metres to be piped to the Negev along the eastern arm of the pipeline.

The third source is sewage water, which after purification will be pumped along the western arm of the Yarkon-Negev pipeline.

The major source, however, will be the Jordan, from which water will flow along huge conduits, which will also collect floodwaters, in the south of the country and in the Negev. The waters will be conserved in reservoirs whence they will flow into the main conduits.

Execution of this project has commenced from the relatively simpler stages, namely the exploitation of the nearer sources, and is proceeding in successive phases.

In 1946/47 an experimental project serving the settlements in the district was launched. These settlements were supplied with approximately one million cubic metres of water from artesian wells in the vicinity. Besides meeting domestic consumption this supply enabled the settlements to begin farming operations. The water was distributed along six-inch pipes. This project, which was not yet regarded as economically sound because of the high price of water pumped through a comparatively expensive and large network, played a role of inestimable importance during the War of Liberation. In spite of the fact that it was a target for enemy operations and sabotage, the project maintained an unbroken supply of water throughout this vital period.

Implementation of the second phase was commenced after the cessation of hostilities and was completed in 1951, when the network of irrigation pipes was extended and 24-inch gauge central pipes and 14–16-inch feeder pipes were installed. In 1952/53 13·5 million cubic metres of water were supplied. This was supplemented by small quantities of saline water with which the fresh water was mixed. The area under irrigation in that year was 27,298 dunams. The water flows through a complex of pipes hundreds of kilometres in length, which is served by a series of pumping stations and of reservoirs with a total capacity of 50,000 cubic metres. These installations are integrated into the central irrigation grid.

The goal of the third phase of the plan is to enable the Negev Authority to supply 125 million cubic metres of water, and involves the construction of the eastern arm of the Yarkon-Negev project. This phase will be constructed in three stages. The first envisages the completion of the Shuval pipeline which is fed by artesian borings in the neighbourhood of Yad Mordechai, and other borings to be executed in the Beersheba District. The first borehole—503 metres in depth—sunk in the Beersheba District, produced 115 cubic metres per hour during tests. An increased flow is expected when the proper equipment is installed. These projects are now being implemented, and when completed the supply of water to the area will aggregate 30 million cubic metres annually.

Stage II, the execution of which was launched in 1953, includes the sinking of additional boreholes in the northern and eastern districts, and the construction of a feeder pipeline from the Gvar Am area. This line, together with the Shuval pipe, will pump water to the terminal section of the Yarkon-Negev 66-inch diameter pipe, the construction of which has begun from the southern terminal. The water will be stored in the central reservoirs at Tekuma, from which it will be pumped to the settlements.

Completion of this stage will raise the annual water supply to 50 million cubic metres.

Stage III involves the pumping of some of the Yarkon water, bringing the gross volume of water supplied up to 125 million cubic metres annually, enabling each settlement to irrigate 2,000 dunams of land—the optimum unit for farming under irrigation.

The three stages of this phase of the project are at present under construction and are expected to be completed within three or four years. Thus sixty villages in the Negev will be assured of a regular supply of water for farm development, while the establishment of a number of other settlements will become possible. Further settlement of the Negev will depend upon execution of other sections of the water-project—completion of the eastern arm of the Yarkon scheme, and the construction of the western arm and the Jordan project.

The three observation stations manned by some dozens of young pioneers, founded a decade ago, were the precursors of thirty-three 'moshavim' and twenty-seven 'kibbutzim' with a total population of 12,738, which dot this arid region today. The town of Beersheba already boasts 20,000 inhabitants. In addition there are three large farming estates, an agricultural school, an experimental station and nurseries for fruit and forest trees.

In 1953/54 900,000 dunams were under cultivation in the Negev, of which 265,000 dunams were cultivated by the Negev settlements themselves, 385,000 dunams by settlements in the northern part of Israel, and by private companies, and 250,000 dunams by Bedouin. With the aid of supplementary irrigation the previous average yield of 30 kilograms of wheat or barley to the dunam has been raised to 150 kilograms to the dunam.

In that year livestock in the Negev included 3,198 head of cattle, producing 6·5 million litres of milk; 9,766 sheep; 196,349 hens, laying 11·5 million eggs. Vineyards and orchards of olives, plums and dates, covering a total area of 3,000 dunams, are beginning to adorn the countryside, in addition to green fields of potatoes, tobacco, tomatoes, onions and flowers.

The obstacles still to be surmounted before this desert region is mastered are legion. Much water is required, as well as a knowledge of more efficient and more economical methods of using water. Closer study of agricultural methods in the Negev, which is still, virtually, terra incognita, is also necessary. There are problems of organization and security to be overcome. But given stamina, vigilance and proper encouragement none of these problems is insoluble.

With regard to the Negev proper, the Negev Plateau and Highlands and the Arraba, in the absence of any plan ensuring a supply of water, the development of any sort of agriculture cannot even be discussed. The few experiments that have already been conducted prove that given water the soil can nourish light vegetation and even shrubs and trees, such as the dalbergia sissoo, Cordia myxa, tamarix and varieties of cactus, such as are grown in the Elath Botanical Garden. In this garden, thanks to irrigation water and the painstaking efforts of a devoted band of workers, whose cherished aspiration it is to fructify the Arraba, fine trees have developed.

Colonization efforts in the Negev must concentrate entirely upon industrial and urban development. Since the occupation of Elath and its resettlement by ex-servicemen, persistent prospecting has been continuing for iron and copper deposits known for the past three thousand years to be available here. Some of these resources have already been discovered but the question whether exploitation is a feasible economic proposition is still under consideration. Meantime phosphate rock is being open-mined, the daily output of 100 tons being utilized for the production of fertilizers. High-grade kaolin, which is being used by the ceramics industry, is being produced, while experiments are being conducted to establish whether the manganese ore, feldspar, mica and other mineral deposits warrant commercial exploitation. This work is being sponsored by the Government-owned Israel Mining Corporation.

Sodom, at the southern tip of the Dead Sea, which lies close to the centre of the Arraba, has revived following the renewed production of potash, bromide, etc. Prospecting for oil is also continuing in this area.

When all these beginnings bear fruit, when the natural resources of the Negev are laid bare, when the highway that has already been constructed is reinforced by a railroad, and Israel's Red Sea port will be crowded with men and ships, then the agricultural hinterland on the borders of the Negev will supply such foodstuffs as the smallholdings, which will be developed here, will be unable to produce for themselves, and then perhaps the genius of Man will discover some way of 'making the dry land into a source of water'.

Settlement plans of such immense proportions involving further agricultural development and expansion imply a superhuman effort that must be made by the Jewish people, the State of Israel and the Settlement authorities.

The Jewish Agency, which is still responsible for new settlement work, is mobilizing tens of millions of pounds for this purpose every year. It has provided the necessary equipment required by the new settlements, as well as by those founded prior to the establishment of the State, but which were not yet fully consolidated.

The volume of these investments is given in the following table (in millions of Israeli pounds):

TABLE 69

Jewish Agency's investments in agriculture

Year	Authorized (£I.)	Spent (£I.)
1947/48	3·5	2·5
1948/49	10·7	7·2
1949/50	16·5	16·7
1950/51	21·8	17·2
1951/52	31·4	27·8
1952/53	40·4	38·0
1953/54	60·0	61·2

In view of the changes in exchange rates and the value of currency at the different periods, it is not possible to make a comparative

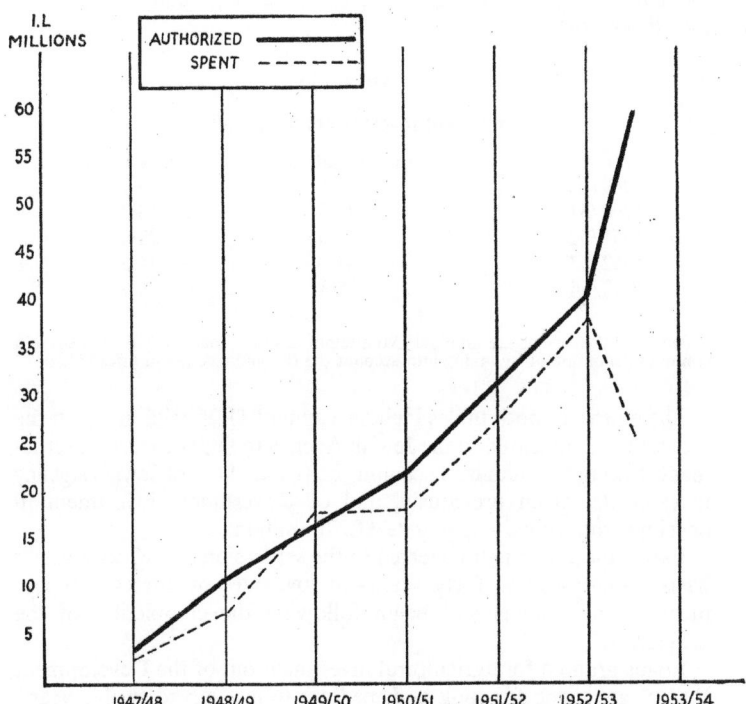

Fig. 8.—Investments in Agriculture of Millions of Israeli Pounds
(1947/48–1953/54)

177

summary of the whole period. We shall rest content with the overall picture rendered, underlining the immense effort made to develop agricultural settlement.

We shall now examine the financial contribution of the Government of Israel with the aid of funds raised within the country, campaigns, foreign loans, grants-in-aid and reparations. In part these were placed at the disposal of the Jewish Agency but mainly they were transmitted direct to the major irrigation and agricultural development projects, at first through various banks, and, for the past three years, through the Israel Agricultural Bank set up in 1951 for the specific purpose of serving the agricultural development of the country.

All sums allocated by the Government for settlement and agricultural development are included in the Development Budget approved every year by the Israeli Knesset. The sums, allocated over the past five years, are given in the following table (millions of Israeli pounds).

TABLE 70

Government investments in agriculture

Year	Amount budgeted	Amount spent
1949/50	11·1	9·9
1950/51	13·3	13·0
1951/52	25·5	22·6
1952/53	41·0	34·5
1953/54	55·3	50·8

Note: All these sums are nominal. No attempt has been made to reduce them to a common denominator by taking into account the rise and changes in price levels.

The figures of both tables include a sum of £I.25 million which the Government allocated to the Jewish Agency to finance work of settlement. Taking this sum into account the two tables—of Jewish Agency investments, given previously, and of Government investment in development, above—aggregate £I.276 million.

Farm equipment is transferred to the settlers on a long-term credit basis—twenty-five to forty years—at low rates of interest. Repayment of these loans will begin following the completion of the settlement.

Loans granted for agricultural investment out of the Development Budget are made for long and medium terms—up to twelve years, but mainly for eight years—through the banks and principally through the Israel Agricultural Bank.

Working capital to finance seasonal, current production comes

178

from all kinds of financial institutions. Such loans are normally made on a short-term basis, usually for twelve months.

While agricultural credit is highly developed in this country its organization leaves much to be desired, both in respect of the borrowers and lenders.

Improvements that appear to us to be desirable for the better organization of this type of credit are as follows:

(1) Lenders must determine the purposes for which the credit is utilized, and must make the money available to borrowers in convenient instalments. Farmers should only make investments if the funds come from suitable sources and conditions of credit meet the needs of the farm.

(2) A system under which each settlement is connected with one financial institution, capable of meeting all demands for credit on the part of the former, is also desirable.

(3) Co-ordination between diverse institutions for agricultural credit is necessary. It is desirable that an Agricultural Credit Authority be established under the aegis of the Government.

(4) Production should be planned with the aid of bulk purchase contracts between marketing agencies and producers under the supervision, and with the encouragement, of the Government. The introduction of this system would facilitate the granting of credits and would improve organized marketing methods.

(5) The provision of means of farm production could be organized on a national and regional basis, thereby obviating the need for each individual settlement to prepare, store and finance stocks.

(6) The import of agricultural equipment on behalf of the Government and the Settlement authorities should be performed through a single public agency which should undertake both the organization and the financing of such imports with a view to directing investments in channels in keeping with settlement and farm development plans.

INVESTMENTS

The development of Israel's agriculture has been achieved by large investments in irrigation works, in the establishment of new farms in all regions of the country and in the expansion of production on old-established farms. Agriculture's share in total gross investments during the six-year period 1949–54 amounts to about one-fourth, as may be seen from the following table overleaf.

The figures in Table 71 for agricultural investments do not include investments in housing for the new farmers. If investments in housing on farms are added the share of agriculture in total investments approaches 28%.

TABLE 71

Total gross investments and investments in agriculture, 1949–54 (at current prices)

Year	Total gross investments (£I. (millions))	Gross investment in agriculture (£I. (millions))	%
1949	92·0	20·0	21·7
1950	132·0	26·0	19·7
1951	188·1	25·1	13·3
1952	275·9	54·5	19·7
1953	300·1	94·7	31·6
1954	440·5	124·5	28·3
Total 1949–54	1,428·6	344·8	24·1

When comparing the above figures it should be taken into account that they represent investments at current prices of each year and that the total, because of the steady and substantial increase in prices, is meaningless apart from the result of agriculture's share during the whole period. Gross investments in agriculture since the establishment of the State (including housing on farms), calculated at 1954 prices, are estimated roughly at £I.1,000 million.

10

THE ROLE OF AGRICULTURE
IN THE NATIONAL ECONOMY

POPULATION

THE population of Palestine at the close of 1947—when the fate
of the country was being decided by the United Nations—
totalled 1,949,000, of whom 630,000 (32·3%) were Jews and
1,319,000 (67·7%) non-Jews. The area of the country was 26,320
square kilometres (excluding 704 kilometres of inland waters), and
the density of population, 74 persons to the square kilometre.

Six years later, at the end of 1953, the population was 1,669,000
persons, of whom 1,483,475 (89%) were Jews and 185,943 (11%)
non-Jews, occupying a total area of 20,406 square kilometres (excluding lakes covering 444 square kilometres). The density of population had risen to 81 persons to the square kilometre. (For purposes
of comparison it is of interest to recall that in 1922 the density was
28 persons to the square kilometre.)[1]

Within and in the immediate vicinity of the three major cities,
population density is, of course, far higher. In the Tel-Aviv area
(the Yarkon Bloc), for example, density at the close of 1952 was
3,048 per square kilometre, in Haifa Bay 869 and in Jerusalem and
the Judean Mountains district—562 per square kilometre. In contrast, the Negev is very sparsely settled, with no more than three
persons to the square kilometre.

[1] Data in this chapter concerning population and its distribution in the
urban and rural districts are: (a) A. Gertz, *Jewish Agriculture in Figures*,
Jerusalem, 1945. (b) *Statistical Handbook of Jewish Palestine 1947*, Department of Statistics of the Jewish Agency, Jerusalem. (c) *Statistical Bulletins
of Israel* edited by the Central Bureau of Statistics.

181

The distribution of population in the urban and rural districts at the close of 1954 is shown in the following table.

TABLE 72

Urban and rural population (%)

	Urban centres	Rural areas
1954		
All Israel	70	30
Jews	76	24
Arabs	27	73
In 1942		
All Palestine	46	54
Jews	76	24
Arabs	27	73
In 1922		
All Palestine	35	65

Despite the metamorphosis set in motion by the War of Liberation, it will be seen that the Arabs in Israel have retained their predominantly rural character. Shifts in the distribution of population over the past three decades, reflected in the growth in the proportion of the number of inhabitants of the urban centres from 35% to 70%, must be attributed to the special features of the Jewish economy. Changes in the political boundaries of the country have exercised only a minor influence in this direction.

The following table indicates the changing pattern of the population distribution over the first five years of the State.

TABLE 73

Growth of Jewish population
(Basis 8.11.48 = 100)

	8.11.48	31.12.49	31.12.50	31.12.51	31.12.52	1953	1954
Total (excluding immigrant hostels)	100	132	165	196	205	215	222
Urban population	100	133	163	185	193	197	201
Rural population	100	146	203	281	299	313	327

The above table shows that since the first National Census (8th November 1948), the total population has more than doubled, with a somewhat smaller increase in the towns and a three-fold increase in the country districts.

Rural population

Ancient Palestine was overwhelmingly agricultural in character, though it included many towns, and even larger cities such as Jerusalem, Hebron, Beersheba, Jaffa, Caesarea, Acre, Lydda, Sephoris, Tiberias, Beth-Yerach, etc. The town-dwellers engaged in commerce, industry and handicrafts. Many of the cities gained renown for specific manufactures. Lydda was a centre of the textile dyeing industry, Jaffa, Acre and Tiberias subsisted to a large extent on fishing. In other centres flax was spun and woven, and mats were manufactured from local reeds. Husbandry, however, continued to be the main

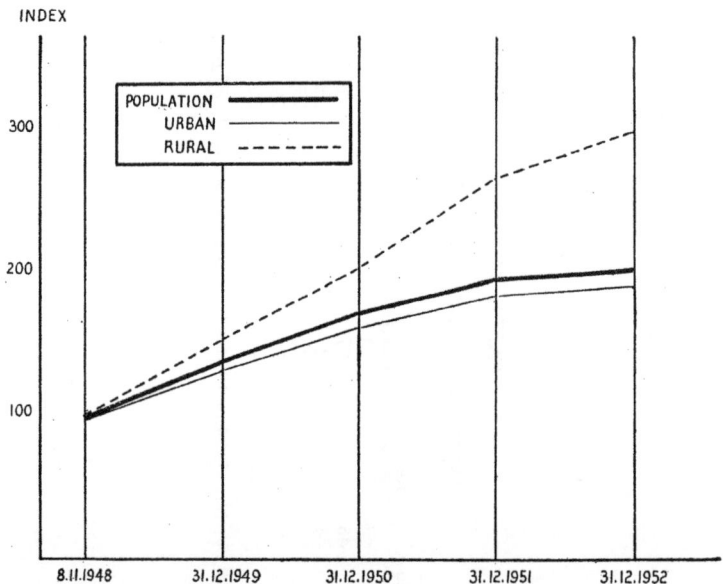

Fig. 9.—Growth of Jewish Population (Basis: 8 November 1948 = 100)

occupation of the nation throughout the era of the monarchy, and till long after the destruction of the Temple in 70 C.E. After the conquest of the country by the Arabs in 632 C.E. the remnants of the Jewish population concentrated in the towns, where they formed communities of varying size.

Jewish immigrants, whose motivation in coming to the country since the Middle Ages was religious, gravitated towards the four Holy Cities—Jerusalem, Hebron, Safad and Tiberias. Prior to the emergence of Political Zionism, in the 'Chibat Zion' period, a handful of idealists envisioned the return to Zion by way of a return to

TABLE 74

Types of settlements in 1922–54

Year	Total no. of settlements	Towns and urban centres	Urban 'moshavot'	Rural 'moshavot'	'Moshavim'	'Moshvei ovdim'	'Moshavim shitufiim'	'Kibbutzim'	Other villages	'Maabarot' and immigrants' hostels
1922	79	8	34		3	8	—	19	7	—
1936	199	27	46		27	43	1	47	8	—
15/5/48	323	28	44		35	59	6	115	36	—
31/12/52	714	39	6	28	40	234	27	217	82	41
31/12/53[1]	744	40	6	29	36	259	29	227	76	42
31/12/54[1]	758	40	6	29	40	259	25	223	106	30

[1] (a) *Statistical Monthly of Israel*, Vol. V, No. 5, May 1954. (b) *Statistical Monthly of Israel*, Vol. VI, No. 3, March 1955.

cultivation of the soil. Indeed the first Jewish agricultural settlement of the modern period, Mikveh Israel, was founded as a farm school. The First 'Aliyah' (wave of immigration), in the concluding decades of the nineteenth century, established no urban centres. In addition to the Mikveh Israel Agricultural College, fifteen other settlements were founded before the end of the century: Petach-Tikvah in 1878; Rishon-Lezion, Rosh-Pinah, Zichron-Yaacov, Ness-Ziona, Yesod-Hamaalah, Gedera and Ekron in 1882–84; Rehovot, Mishmar-Hayarden, Hadera, Motza, Hartuv, Beertuvia and Metullah in 1890–96.

The launching of the Zionist Movement gave a new impetus to the influx of immigrants, and during the Second 'Aliyah,' in the early 1900's, many new settlements, both rural and urban, were founded. A new Jerusalem developed beyond the ramparts of the Old City and near ancient Jaffa the town of Tel-Aviv arose. In the immediate vicinity of Haifa, Tiberias and Safed Jewish towns were founded. In this period before the outbreak of the First World War another nine villages were established. The work of settlement was renewed some years after the cessation of hostilities, and from 1921 onwards a comparatively large number of new settlements and towns were established.

The cities and urban centres include six towns with a mixed population. The category 'other villages' included (in 1954) 20 labour villages, 27 rural settlements and 59 educational institutions. There were also in that year 27 autonomous 'maabarot', with a total population of 39,196, and 6,342 residents in immigrant hostels.

The total population in 1900 of the 15 villages then in existence was 2,811. In 1914 there were 24 Jewish villages with a total of 10,889 inhabitants.

According to the first National Census in 1948 the Jewish population of Israel totalled 715,940, of whom 530,433 were resident in the towns. Relevant information regarding 29,816 persons is lacking; these were mainly new immigrants. 155,691 Jews, or 22·6% of those whose permanent place of residence was known (21·7% of the total population), were domiciled in the villages.

For purposes of comparison it is of interest to cite some population figures for the United States of America. In the year 1880 71·4% of the population were resident in the rural districts; by 1940 this proportion had declined to 43·5%. In the course of this sixty-year interval the population of the United States grew by 162·5%. The growth of the rural population, however, was only 59·6%.[1] The census of 1950 revealed that the population had increased by 14·4%

[1] U.S. Census Report 1880–1940.

THE ROLE OF AGRICULTURE IN THE NATIONAL ECONOMY

since 1940,[1] but the rural population had declined by 4·5 million persons.[2]

Rural population, of course, is not to be confused with agricultural population, as a large proportion of the former engages in non-agricultural occupations. In the United States the rural population category includes mining villages, suburbs of the cities, etc. In 1940 27,094,047 persons not engaged in agriculture—20·6% of the total population, or 47·3% of the rural population—were domiciled in the villages.

The distribution of the population of Israel at the end of 1954, classified according to the types of settlement, was as follows:

TABLE 75

Distribution of the population of Israel

	No. of settlements	Population	%
Total	862	1,717,834	100·0
Towns	21	1,036,877†	60·4
Urban settlements*	21	103,592	6·0
Bedouin tribes	—	20,445	1·2
Urban 'moshavot'	6	70,300	4·1
Rural 'moshavot' and private 'moshavim'	69	78,550	4·6
'Moshvei ovdim'	259	89,383	5·2
'Moshavim shitufiim'	25	4,455	0·3
'Kibbutzim'	223	76,115	4·4
Labour villages	20	4,271	0·2
Autonomous 'maabarot'	27	39,196	2·3
Other rural settlements	27	63,184	3·7
Immigrants' hostels	3	4,342	0·2
Farms and farm schools	59	6,628	0·4
Arab villages	102	120,496	7·0

* Includes 3 labour camps.   † Includes 50,877 non-Jews.

In order to arrive at a more accurate estimate of the urban population, it is necessary to add the number of residents of other centres of an urban character, such as labour camps, urban 'moshavot' and at least half of the inmates of the immigrants' hostels and the 'maabarot' situated in the immediate vicinity of the towns. Thus at the close of 1954 the distribution of the total population of the country was:

Urban centres 70·5%
Rural centres 29·5%

---

[1] F.A.O. Reports, Vol. V, No. 1, 1950, p. 26.

[2] J. D. Black, *Introduction to Economics for Agriculture*, New York, 1953, p. 860.

For the Jewish population alone, however, the distribution was:

Urban centres 76·2%
Rural centres 23·8%

In the United States of America the decline in the rural population is a corollary of the expansion of industry, the mechanization of agriculture and rising levels of farm production. In the five-year period 1948–53 the distribution of the population of Israel between the towns and country districts did not change to any significant extent despite an aggregate increase in population of 107%, and the constant movement from the country districts to the towns, normally intensified in a period of mass immigration.

AGRICULTURAL OCCUPATIONS

But the distribution of population in urban and rural districts does not reflect the place of agriculture in the national economy. For this purpose closer study is necessary of the occupational structure of the State, and of the economic role of agriculture in terms of production and consumption, exports and imports, share in the national income and the balance of payments.

TABLE 76

Changes in the occupational structure of the population

	1942		1954†	
	% of earners		% of earners	
	Of total population	Of Jews only*	Of total population	Of Jews only
(A) Primary occupations (agriculture, afforestation, fishing)	46·7	13·2	18·2	14·9
(B) Secondary occupations (industry, trades, electrical power, public works and building)	22·4	38·2	33·8	35·0
(C) Tertiary occupations (commerce, transport, public and other services, liberal professions, administration, etc.)	30·9	48·6	48·0	50·1
	100·0	100·0	100·0	100·0
Number of earners	595,000	212,000	561,400	518,500

* Figures for the Jewish population refer to 1943.
† Labour Force Survey, June 1954.

According to the results of the Labour Force Survey, carried out in June 1954, the number of earners among the total population was about 560,000, or only one-third of the total population. The share of earners in the Jewish population was, according to the same survey, 35%, as compared with about 40% before 1948. But it should be taken into account that the Labour Force Survey in June 1954 was based on a population sample of 14,790 families and that the results may slightly differ from the real numbers owing to 'sample errors' and inaccuracies in replies. It is held that the total number of earners represents an underestimate and the real number of earners may amount to about 600,000, constituting approximately 36% of the total population. The results of the survey reflecting the occupational structure of the population are held to be reliable.

A comparison of changes in area, density and the number of earners over the past seven years will prove instructive.

TABLE 77

Comparative changes in area, density and earners

Year	Area	Density	Earners
1947	100	100	100
1954	77	115	100

It will be seen from the foregoing table that the area declined by 23%, density increased by 15% and the number of earners remained constant. In this period, moreover, the economy of the country was revolutionized. Palestine was a mainly agrarian country with 58·8% of the population—according to the Census of 1931—engaged in agriculture. By 1942 this figure had declined to 46·7%, not changing to any material extent down to 1947. This decline was reflected in the Jewish sector of the population by a decrease in the proportion of those engaged in agriculture from 13·2% to 12·6%.

The economic metamorphosis was a direct result of the change in political conditions. Israel is no longer an agrarian country, and is not likely to become one even if agriculture should expand and employ more than the present 18% of the total number of earners, and achieve the dimensions necessary for the domestic production of its foodstuffs.

Before discussing the maximum possible contribution of agriculture to the national economy it is instructive to note that the ratio between the three main groups of earners has changed from 13 : 38 : 49 twelve years ago to 15 : 35 : 50 today. The first—agriculture, afforestation, fishing—category has increased at the expense of the second—industry, trades, building, public works, etc.

The initial five years in the life of a young state do not provide an adequate basis for any far-reaching conclusions. Nevertheless these trends in the occupational structure of the population appear to indicate a definite process shaping Israel's economy. It is highly probable that further changes will be registered in the future in the relative weight of these three categories, and that the primary and secondary groups will increase at the expense of the third. We shall endeavour to arrive at a reliable estimate of the optimum proportions of agriculture in a stable national economy.

Some two-thirds of the world's population today derive their livelihood from agriculture. The remaining third comprises 14% engaged in industry, 6% in trade and 13% in transport, services and other occupations.

Only in seven countries—Great Britain, Australia, the United States, Belgium, Germany, Holland and Switzerland—is the proportion engaged in agriculture lower than that in industry.

According to one study[1] the proportion of the total number of earners in various countries subsisting on agriculture was as follows in 1930:

Europe (excluding Russia)	37%
Russia	87%
North America	23%
Central and South America	65%
Africa	75%
Asia	75%
Ukraine	30%

A comparison of Israel's occupational structure with that of two countries producing a surplus of agricultural produce for export, United States and Holland—and two others which import foodstuffs from abroad—England and Belgium, is given in the table that follows.

TABLE 78

Occupational groups in different countries (%)

Category	U.S.A. 1940	Holland 1947	England 1950	Belgium 1947	Israel 1952
A	17·6	20·2	5·2	12·3	14·7
B	32·2	33·7	49·2	49·2	29·8
C	50·2	46·1	45·6	38·5	55·5

We have not included any agrarian countries in this table. We have chosen two highly industrialized countries like England and Belgium, and two countries, the economic structure of which lies between the two extremes.

This table indicates that in Israel the percentage of persons engaged in the tertiary occupations is larger than in any of the other

[1] *World Agricultural Survey*, London, 1952.

countries—and far in excess, of course, than in countries with a predominantly agrarian economy. In Turkey, for example, the third category accounts for only 15·1% of the total population, in Japan—24·6%, in Egypt—19·3%. Naturally in countries with higher standards of civilization, public services are far more highly developed. In Israel the concentration in this category is excessive. Among the causes we may cite the necessity to fill with Jews all the positions in the national and the local government administration rendered vacant after the departure of the British, and—to lesser extent—the Arabs. This category also includes the Regular Army. Another reason is the relative ease with which new immigrants are absorbed in petty trading and services, as compared with agriculture and industry. Absorption in the two first categories is more difficult because of the higher standards of training and skill required, as well as the need for investment capital—far more than in peddling, guard duties, services and the like.

However, a breakdown of the number of Jewish earners gives a rather more encouraging picture, reflected in the substantial quantitative increase in all categories.

TABLE 79

Earners—numbers and indices

	1947	1952	Index 1947 = 100
(A) Agriculture	32,000	78,000	244
(B) Industry, etc.	89,000	168,000	178
(C) Commerce, services, etc.	132,000	294,000	223
	253,000	540,000	209
Earners as percentage of population	40	36	

Current estimates of the number and composition of the earners in this country must be accepted with reserve. We believe that estimates of the number of those engaged in agriculture are exaggerated because of excessive weight given to the work of women in agriculture without adequate allowance being made for their work in housekeeping. Probably a figure of 64,000 earners in agriculture, rather than 78,000, would be more correct for 1952. The difference of 14,000 should be subtracted from the total. It is highly desirable that a new study of the occupational structure be undertaken, but until then we must perforce rest content with the figures available.

In the course of five years the total number of earners has doubled. Of this increase of 277,000 17% were absorbed in agriculture, 25% in industry, etc., and 58% in commerce, services and the like.

The decline in the number of earners to below 40% which is considered the norm, should be noted. Because of the age-structure of the new immigrants Israel lacks 50,000 earners.

In considering this top-heavy occupational structure, we must recall the complaints made by immigrants about inadequate aid to enable them to enter a more productive calling on their arrival in Israel. Even older residents, organized within the Town to Country Movement, failed to achieve their objective because of the delay in the construction of housing and the acquisition of equipment. The main reasons for this disequilibrium, nevertheless, remain the composition and the age-structure of the influx of immigrants, tending towards the higher age-groups—in contrast to previous waves of immigration—and conditions reigning in the period immediately following the Second World War.

Overmuch attention should not be paid to complaints regarding the disparity between execution and planning. In no country has any agricultural plan been implemented as drafted. States which boast of plans realized to an extent varying between 137% and 150% do not differ in this respect from those achieving no more than 60 or 70% of their objectives. No government operates in a void; its work is always dependent upon a wide variety of external and internal factors, not all of which can be foreseen. The slowing down of Israel's irrigation plans in 1950–51, to cite one example, was caused by the outbreak of the Korean War upsetting international trade, and causing an exorbitant rise in the price of steel pipes. Supplies available fell short by far of the demands of ambitious development programmes in various parts of the world. Fund-raising campaigns and other schemes to raise investment capital have also often not come up to expectations.

Our discussion so far has been confined to the occupational distribution of earners in Israel. This alone, however, is not sufficient to explain the country's economic structure. In the United States, 18% of the population produce 90% of the foodstuffs consumed locally, with large surpluses remaining for export. The remaining 10% comprise mainly coffee, sugar and spices.

In Israel 14% of the population engaged in agriculture produce only 50% of the country's consumption. Exports of agricultural produce, mainly citrus, cover no more than a small proportion of the deficit. It is abundantly clear that even an increase in the percentage of the population engaged in agriculture, say from the present 14% to 18% as in the United States, would not produce similar results. The production of so high a proportion of domestic consumption as well as of large surpluses for export is dependent upon a variety of other factors, including productivity, standards of equipment, yields, etc.

In the countries already referred to the decline in the number of earners engaged in agriculture must not be attributed to any reduced demand for farm produce. It is rather the direct result of better agricultural methods, and especially of the rapid pace of industrial expansion.

In the United States in 1880 the rural population comprised 71·4% of the national total. By 1940 it had declined to 43·5%. But exports of agricultural produce rose from 330 million dollars (or 83·5% of all exports) in 1880 to 3,411 million dollars in 1950 (27·1% of total exports).[1]

The declining percentage of earners in the United States engaged in agriculture and the relative increase in industry are shown in the table that follows:

	1910	1930	1940
Agriculture	33·2	21·4	17·6
Industry	27·8	28·9	32·2

The indices for the export of wheat, flour and grains were 1924–29 = 100.

	1930	1950
Wheat and flour	71	204
Other grains	28	193

In absolute figures the agricultural population of the United States —approximately 32 millions—has remained virtually unchanged. The number of farm-units, too, remained comparatively stable for three decades, declining from 6,300,000 units in 1910 to 6,096,000 units in 1940. In the subsequent decade, however, the number of farms dropped to 5,384,000 (in 1950), offset by a 29·3 million hectare increase in the cultivated area. We have already mentioned some of the agricultural commodities imported into the United States, either because local production falls short of consumption—wool, hides, sugar, copra, etc.—or which cannot be produced locally at all, such as tea, coffee, rubber, silk, cocoa, bananas and spices.

Agricultural exports more than cover the cost of agricultural imports by a figure ranging between one and two billion dollars.

Another criterion is the contribution of agriculture towards the national income. In Palestine, according to a study carried out in 1936, agriculture accounted for 16·4% of the national income— 9·4% in the Jewish and 24·2% in the non-Jewish sector.

[1] 'U.S. Farm Products in Foreign Trade', *Statistical Bulletin*, No. 112, Washington, 1953.

In 1940 in Israel the figure was 9·0%, in 1950—9·1%, in 1951—10% and in 1954—13·8%.[1]

In the United States the share of agriculture in the national income was as follows: 1919—17·4%; 1929—9·6%; 1939—8·8% and 1949—8·4%.

In Israel, in the year 1951/52, the value of agricultural production aggregated 96 million dollars, of which produce to the value of 62 millions was intended for direct human consumption, 15 million dollars for feeding stuffs and 19 million were exported (mainly citrus). In that year foodstuffs to a value of 65 million dollars were imported while 30 million dollars were spent on imported materials, including spare parts, fuel, fertilizers, fodder, etc. The value of the local consumption of foodstuffs, accordingly, was 191 million dollars, of which 50% was locally produced.

The relation between domestic and imported produce (including agricultural materials) is reflected in the following table:

TABLE 80

Domestic production of the main nutrients[2]

Daily per person	Calories	Protein (gm.)	Animal protein (gm.)	Fats (gm.)
Consumption	2,706	84·7	26·2	68·8
Domestic production	713	28	14·6	19·6
Domestic production %	26	33	55	28

It is only in the category of animal proteins that domestic production contributes more than half of consumption. Of proteins generally the proportion is one-third and of fats and calories rather more than a quarter.

In view of the acute lack of foreign currency and the over-riding necessity of increasing local production and reducing imports, over-all planning must concentrate upon those branches of agriculture that constitute a source of animal proteins—the most expensive item in the import list—and more specifically upon dairy-farming and marine fishing. Concurrently other branches will be expanded, such as fruit-growing, vegetables and potatoes. All these, together with sugar to be produced from locally grown crops, will constitute a valuable supplement to the consumption of calories. Wheat for bread production will continue to be imported for a long time to come.

[1] (a) *Agricultural Studies*, No. 2, p. 16, published by the Israel Ministry of Finance, December 1950. (b) For 1954—Economic Advisory Staff's estimate.

[2] *Statistical Monthly of Israel*, Vol. V, No. 6, June 1954, p. 370.

We may assume that one-third of the consumption of calories and of the proteins (providing that all animal proteins consumed are produced domestically) respectively will be imported. The supply of fats presents a more difficult problem, in view of the comparatively small area under oil-producing plants. Only if the cultivation of ground-nuts is sufficiently expanded can Israel hope to supply the requisite minimum of fats and oils from domestic production.

The common objective of the competent authorities in the spheres of production, import and supply, is to stimulate home-production and enable a contraction of imports of foodstuffs.

NUTRITION PROBLEMS

Just as the eye instinctively concentrates upon the trunk and foliage of a tree, forgetful of the roots running deep into the soil, so outward expressions of hunger and the fight constantly waged against it divert attention from its causes.

An inquiry into the causes and the prevention of hunger would save untold human lives, would avert wars, and in the ultimate resort would prove far more economical than the production and maintenance of the immense machinery of destruction. A proper study of hunger in India and China, North and South Korea, Vietnam and Vietminh, might well have enabled mankind to avoid the Indo-Chinese and Korean Wars. If instead of supplying arms to Iraq, efforts were directed towards harnessing the Tigris and the Euphrates, the Great Powers would make a far greater contribution towards human happiness and progress.

Human beings need suitable nutrition. Agriculture must be so planned and organized as to be the universal provider.

International institutions, under the aegis of the League of Nations in the past and the United Nations Organization today, have not succeeded in placing the reciprocal relations of the nations upon a more rational basis and avoiding wars. But some of these institutions are engaged in the study of vital problems of the human race and seek to discover methods of improving living conditions. One of the most important of these international bodies is the Food and Agriculture Organization, which has done trojan work in the fields of agriculture and nutrition. The computation of minimum quotas of main nutrients has greatly facilitated the work of agricultural planners.

In recent years an annual balance of foodstuffs has been drawn up in Israel. Comparison of three of these balances indicates changing nutrition standards which are dependent upon both domestic production and import policies.

TABLE 81

Changes in food baskets[1]

	Kilograms per year			
	1949/50	1950/51	1951/52	Optimum
Cereals and products	131·2	133·1	146·6	130
Potatoes and starches	46·3	41·8	46·6	50
Sugar and honey	17·5	19·2	19·7	24
Pulses and oil seeds	8·7	8·5	6·8	8
Vegetables	108·0	102·2	111·2	150
Fruit (incl. cucurbits)	107·6	103·8	124·8	125
Meat	18·9	15·2	10·7	12
Eggs (incl. egg powder)	14·8	15·6	12·8	15
Fish	17·4	18·6	15·6	18
Milk and its products	96·3	96·7	93·6	145
Oils and fats	15·5	17·1	16·0	18
Miscellaneous	8·2	7·7	6·7	8

The average food basket typical of each of the three years under review contained the following quantities of main nutrients.

TABLE 82

Nutrition values per capita according to food baskets[2]

	1949/50	1950/51	1951/52	Optimum (hypothetical)
Calories (daily)	2,613·0	2,681·0	2,706·0	2,690·0
Proteins (daily) (gm.)	83·7	87·7	84·7	87·0 (incl. 30 gm. animal proteins)
Fats (daily) (gm.)	75·0	75·4	68·8	73·0

The substantial rise in the consumption of grains may perhaps be explained by the widespread practice of feeding bread to poultry owing to the restrictions imposed upon the import of feeding stuffs. It is, however, also due to the growing human consumption of grain, reflected in increased intake of calories in these three years, to the hypothetical optimum. In regard to other foodstuffs the rise in consumption is mainly in domestic produce, including vegetables, potatoes, fruit, offset by a decrease in imported foodstuffs, such as oil and oil seeds, meat, eggs (the production of which is dependent upon imported feeding stuffs), fish, milk products (especially cheese) and dried fruits.

[1] *Statistical Monthly of Israel*, Vol. V, No. 6, June 1954, p. 370.

[2] H. Halperin, *Agricultural Production and Mass Immigration in Palestine*, Tel-Aviv, 1953, p. 18.

BREAD: PRODUCTION OR IMPORT

It is generally accepted that Israel must continue to import tea, coffee and spices, for it is doubtful whether, in the near future at least, it will prove possible to cultivate such tropical plants here. That this should also be the case in regard to wheat may give rise to some surprise. In previous chapters we have discussed the lack of land for extensive cultivation of unirrigated crops. Should Israel desire to produce its entire consumption of wheat—300,000 tons annually—at least three million dunams would be required for this purpose and under the most primitive rotation of crops, say the two-year system, it would need double this area. Under present conditions the total cultivable area in the country falls short of four million dunams. Agriculture based upon grain cultivation exists in backward countries possessing a low standard of living. To produce sufficient wheat for local consumption the country would require a cultivable area five times as large as the present, or at least a far larger area under irrigation and the introduction of wheat into the crop-rotation. Proposals have been put forward for the introduction of wheat into a four-year system of crop-rotation, together with one-yearly crops of fodder, vegetables, potatoes and industrial crops; in other words, after three years of irrigated cultivation, the same area would be seeded with unirrigated wheat. The wheat, it is argued, would produce high yields because of the moisture retained in the soil and at the same time would improve the latter by drying it out in preparation for the next three years of irrigated cultivation. This proposal, fathered by an agricultural expert, is worthy of being tested before the implementation of large-scale irrigation plans.

It has been argued that prior to the establishment of the State wheat production was of substantial proportions. It is perfectly true that the Arabs engaged largely in the cultivation of wheat. The total annual yield (in thousands of tons) since 1929 is reflected in the following table (in brackets the yield of Jewish farmers):[1]

1929	1930	1931	1932	1933	1934	1935	1936	1937	1938
87	87	79	51	44	82	104	76	127	44
							(11)		

1939	1940	1941	1942	1943	1944	1945	1946	1947
89	136	90	104	60	57	n.f.*	n.f.*	n.f.*
				(13)		(11)	(16)	(12)

* n.f.—no figures.

---

[1] Figures compiled from different sources: Government Department of Agriculture, the Jewish Agency and Field-crops Growers' Association.

The area under wheat was as follows (in millions of dunams):

1937	1938	1942	1944
2·2	2·1	2·0	1·5

Similar areas were devoted to barley cultivation. In these years grain was grown on 79% of the total area under cultivation in Palestine.

The following table shows the area under winter grains in 1952/53:[1]

	All farms (dunams)	Jewish farms (dunams)
Barley	827,488	617,103
Wheat	346,912	195,649

The two main winter grain crops accounted for 30% of the total cultivated area.

Palestine, under the Mandatory administration, also imported large quantities of wheat and flour, averaging about 130,000 tons yearly. The sources available do not state the precise figures for import of wheat from Transjordan.

In 1953 the import of wheat and flour aggregated 300,000 tons. Plans drawn up for the 1953–60 period envisage annual wheat imports of 175,000 tons.

OPTIMUM DIMENSIONS OF AGRICULTURE

We shall attempt an estimate of the number of earners, out of a total population of two millions, who must engage in agriculture to supply 75%–80% of the food consumption of the country, and to produce a surplus sufficient to finance the import of the balance.

We shall assume the area under cultivation and the number of livestock to be that contemplated in the agricultural development plans for the years 1954–60. Work-norms will be as at present, averaged between those current in collective and individual farms.

Beekeeping must be added to the branches of farming producing food for man and beast; other agricultural branches, such as nurseries, flower-gardening (for both the local and export markets), etc., have been included in the 'miscellaneous' category. These branches account for an estimated 5% of the total amount of labour invested in agriculture.

Assuming that a further 10% of the total number of work-days will be necessary for maintenance of machinery and implements and for tending draught animals, we arrive at a total of 33 million days of

---

[1] *Statistical Abstract of Israel*, 1953/54, No. 5, p. 58.

TABLE 83

Hypothetical balance of work-days per annum in agriculture

Crop	Area (dunams or units)	Work-norms (No. of work-days per dunam)	Total no. of work-days per annum
Vegetables and potatoes under irrigation	230,000	16	3,680,000
Vegetables and potatoes (un-irrigated)	22,000	1	22,000
Sugar-beet	95,000	10	950,000
Ground-nuts	290,000	10	2,900,000
Artificial pastures	72,000	5	360,000
Lucerne	50,000	6	300,000
Irrigated fodder	262,000	4	1,048,000
Unirrigated fodder	310,000	1·5	465,000
Irrigated winter grains	250,000	2·5	625,000
Unirrigated grains	840,000	0·5	420,000
Summer grains	340,000	3	1,020,000
Irrigated legumes	250,000	3	750,000
Unirrigated legumes	26,000	1	26,000
Cucurbits (irrigated)	7,500	4	30,000
Cucurbits (unirrigated)	30,000	3	90,000
Tobacco (irrigated)	4,500	12	54,000
Tobacco (unirrigated)	52,000	16	832,000
Cotton	70,000	12	840,000
Oil seeds	45,000	2	90,000
Flax	10,000	1	10,000
Green manure (irrigated)	260,000	0·75	195,000
Green manure (unirrigated)	35,000	0·2	7,000
Citrus	215,000	20	4,300,000
Orchards (irrigated)	140,000	20	2,800,000
Orchards (unirrigated)	220,000	8	1,760,000
Fish-ponds	40,000	6	240,000
Milch cows (head)	62,000	30	1,984,000
Calves (head)	70,000	16	1,120,000
Sheep (head)	150,000	3	450,000
Laying birds	3,000,000	0·22	660,000
Pullets	1,500,000	0·1	150,000
Miscellaneous (5%)			1,408,000
Total			29,586,000

labour. The farmer's working year comprises 264 days, after sub-tracting 52 Saturdays, 12 holidays, 14 days of vacation, an estimated 7 days of illness and 16 days made up by the shorter working day on Fridays. Thus it is computed that 125,000 persons must engage directly in agriculture—exclusive of ancillary occupations. Fisheries and afforestation, of course, also belong to the primary occupations. Out of a total population of two millions, 740,000 persons, or 37%, will be gainfully employed. In view of the preponderance of the eastern communities in the population, among whom the proportion of large families is high, we do not foresee any possibility of achieving the required 40% of earners in this period. We may predict therefore that more than 18% of the earners will engage in agriculture, producing approximately 75% of the country's food consumption. In the villages 25% of the population must engage in non-agricultural professions and trades. Basing our calculations upon an average of four persons to the family, an estimated 660,000 persons, or 33% of the total population, will live in the country districts.

In our study of the distribution of the urban and rural population in Israel we found that 26·2% of the aggregate and 21·2% of the Jews were resident in the villages. The rise of the average to 33%, as indicated above, implies an increase of the Jewish rural population to 28% and in the number of earners engaged in agriculture from 11% to 18%.

The optimum contribution of Israel's agriculture to the national economy will be determined by the consumption of foodstuffs, the rate and dimensions of development, technological standards and the skill of the farmers. Demand is determined by tastes and purchasing power and its dimensions are governed by real income per capita of the population. The progress of agriculture will be largely dependent upon purchasing power created in other branches of the economy; the latter, reciprocally, will depend upon agriculture. Even in highly industrialized countries, efforts are constantly being made to stimulate agricultural production, by technological improvements, by increasing the fertility of the soil and by raising the living standards of the farming population.

Britain, which for six generations has consistently pursued a policy of industrial development, paying for cheap foodstuffs—today no longer cheap—with its manufactures, is now seeking to develop its agriculture, is actively encouraging a 'Town to Country' movement, has adopted suitable agricultural price policies, is planning production and marketing—all with the object of enhancing the security of the farming community.

Natural conditions in this country require maximum intensification of agriculture, the construction of large irrigation works and the introduction of modern farm machinery and equipment in all branches. Above all high standards of skill and 'know-how' are essential.

From our experience in the past we know that in times of stress and emergency Israel's farming economy constitutes not only the national larder but a reserve of manpower upon which the country can rely in any emergency. The need to unite in the farmer such diverse functions as entrepreneur, planner, director and executor, brings his personality to maximum development and stability.

Such farmers, on a high cultural level, backed by a nation of an equal level, will mould the character, and the quality and scope of Israel's agriculture.

11

RURAL SOCIOLOGY IN ISRAEL

C O-OPERATIVE organization was completely unknown in Palestine in the late nineteenth century, when Jewish colonization of the country began. The social and economic order in the Arab village, hallowed by a tradition of generations, was, and for the most part still is, immune to all change. During the period of the Mandatory administration the persistent attempts to introduce the elements of co-operative organization fell on stony ground, and later the efforts of the Israel Government in the same direction have been equally unfruitful. In one district, for example, where the Government had sunk an artesian well, opposition of the villagers effectively blocked all endeavours to organize the water supply upon a co-operative basis. Ultimately the authorities were compelled to establish a public company managed by a Government official. Another instance is that the credit co-operative society which serves the Arab population has had difficulty in finding personnel from its own members capable of running the society and has required to turn to outside assistance.

Jewish settlers who fifty years ago organized co-operative credit and citrus marketing societies were pioneers in Palestine. Their experiments were based upon overseas models.

Development was rapid and the co-operative form of organization soon penetrated into a number of other fields. This development was marked in agriculture, where soon a ramified network of co-operative creameries, packing houses, refrigeration plants, seed-raising farms, etc., was created. The first villages founded by Jewish settlers—the 'moshavot'—from the 1870's onwards were based on the family unit and hired labour (in certain cases the settlers did not even live on their farms). In this respect they hardly differed from the type of village common in many other countries. The establishment of villages upon a new co-operative basis was not undertaken until

201

the first decade of the present century when settlers, fired by the ideal of living a life of labour in Palestine, entered the country. These settlers developed an extensive co-operative movement, which by virtue both of its achievements and its specific character, has made a notable contribution towards the promotion of the principles of co-operation throughout the world.

The workers' agricultural co-operative movement constitutes a distinctive section of the co-operative movement in Israel, because of its characteristic structure and objectives. Its basic principles were evolved within the Palestine Labour Movement, of which it is an integral part. Its motive force is the ideal to establish a new society in Palestine upon the foundations of social justice. The twin tenets which have guided the co-operative movement until today are: the evolution of a Jewish agricultural economy, directed towards a maximum degree of self-sufficiency, in order to serve as the basis of a national economy; and scrupulous observance of the principle of self-labour to obviate any social exploitation. The workers' agricultural co-operative movement is one of the most striking features of the co-operative movement in Israel. Agricultural co-operation in this country covers practically every field. Of cardinal importance, of course, is agricultural production, but because its essential role was colonization the co-operative movement has of necessity engaged in the organization of marketing, purchase of supplies and equipment, financing of settlement, irrigation works, the production of seeds and the like. Because the workers have been concerned to develop agriculture in general they have also interested themselves in the cultivation of privately owned orchards and groves. To this end, they have organized co-operative labour contracting agencies and housing projects for hired labourers.

Both in form of organization and objectives the workers' agricultural co-operatives in Israel differ from similar groups in other countries. Neither the 'kibbutz', based upon communal production and consumption, nor the 'moshav ovdim', the co-operative smallholders' settlement, possess any counterpart elsewhere; and indeed, even in Israel, they constitute a unique section of the community.

Both the 'kibbutz' and the 'moshav' are agricultural co-operative societies, though of an unusual character. In the 'kibbutz' and the 'kvutza' co-operation has assumed radical collectivist forms, such as the complete absence of private property and absolute equality among all members. In the 'moshav ovdim' co-operative principles are limited to specific spheres, such as the sale of farm produce, the purchase of supplies and equipment, the provision of credit, the operation of heavy agricultural machinery, etc. The distribution of profits individually among the members, which is common in other co-operatives, is absent both in the 'kibbutz' and the 'moshav'.

The economic and social principles to which both the 'kibbutz' and the 'moshav' subscribe are: (*a*) settlement on nationally or publicly owned land; (*b*) self-labour of the settlers, to avoid any possibility of exploitation of hired labour; (*c*) mutual aid, and (*d*) co-operative purchase and sale.

The differences between the 'kibbutz' and the 'moshav' are fundamental. The 'kibbutz' believes in communal ownership of all means of production and consumption and insists upon the full equality of rights and obligations of all of its members. In the 'moshav ovdim' each settler cultivates his farm on an individual basis, on his own account and responsibility. Over the past twenty years a new type of 'moshav ovdim' has developed—the 'moshav shitufi', in which most branches of the farm are operated upon a collective basis, but consumption is individual, each family preparing its own food and taking its meals privately and not in a communal dining-hall as in the 'kibbutz'. In the early stages of the 'moshav shitufi' varying degrees of co-operation characterized different villages of this type. Thus there were villages in which *either* field crops *or* fruit orchards were cultivated on collective lines; others in which *both* of these branches were cultivated collectively, members working smallholdings near their own cottages, with or without the approval of the village committee.

Comparisons are frequently made between agricultural co-operative settlement in Israel and the planned development of the agricultural economy in the Soviet Union. The collective character of production, the collective ownership of means of production, the peculiar way of life and collective principles of consumption, all these characteristics which are typical of the 'kibbutzim' are compared with similar features of villages which they are supposed to resemble in Russia—the 'kolkhoz' and the 'sovkhoz'. The comparison, however, is artificial and applies only in regard to external features. It is a matter of common knowledge that the foundation of the 'kibbutz' predates that of the first 'kolkhozes' and 'sovkhozes' by a number of years.

The 'sovkhoz', of course, is no more than a government farm, employing hired labour and managed by state officials. There is, accordingly, no basis for comparing the 'sovkhoz' with an independent farm, whether run on collective or individual lines. The employee of the 'sovkhoz' is not a producer seeking to supply his own needs; he is rather a worker in a gigantic agricultural factory, producing for the market. The labourer on the 'sovkhoz' is hired on a fixed wage basis like his counterpart in any urban industry.

The 'kolkhoz' has developed into three subtypes. Only its higher type which is already obsolete was based upon collectivist principles and could therefore be compared even with the 'kibbutz'. Essentially, however, it had little in common with the latter. The 'kolkhoz' was evolved from the very outset as a political instrument in the struggle against the individualist Russian villager. The 'kibbutz', it cannot be too strongly stressed, aspires of its own free will and not because of any extraneous State compulsion—as is the case with the 'kolkhoz'— to create a collective way of life realizing collective principles.

In addition to this fundamental dissimilarity, new regulations which have been introduced into the administration of the 'kolkhoz' provide for the computation of earnings of the workers upon the basis of the number of days worked and aggregate production, and also allow every member a small privately owned plot for domestic cultivation. The disparity between these two types of agricultural organization, it is clear, is vast.

To some extent the 'moshav shitufi' may be compared with the 'kolkhoz', but here too the likeness is more apparent than real, as it does not take into account the respective degree of liberty and private initiative enjoyed by the farmers in the two countries.

Taking economic, organizational and social development as a criterion, villages in Israel can be classified into a number of categories:

(1) The *Arab village* which only after the establishment of the State of Israel rid itself of the final traces of feudal character; today most Arab farms belong to the small or medium family farm class.

(2) The '*Moshava*', most or some of whose inhabitants subsist on agriculture. The farms, which are cultivated with the aid of hired labour, are for the most part monocultural in character, concentrating upon citrus groves or vineyards. In many cases the owner of the farm does not live in the village and may even engage in a non-agricultural occupation.

(3) The '*Moshav Ovdim*', based upon individual labour and way of life. The settlers of this type of village support themselves by farming their holdings with the aid of the members of their families. Their way of life is governed by prescribed social principles, among the most important of which are: national ownership of the soil, self-labour, mutual aid and co-operative marketing of produce and purchase of supplies and equipment.

(4) The '*Moshav Shitufi*'. In this type of village consumption is individual but agricultural production is on a collective basis (the degree of which may vary from village to village). In some of these 'moshavim' all branches of farming are conducted collectively, in others this is the case only in the major agricultural branches.

(5) *'Irgun Lehityashvut'*,[1] is the name given to a co-operative group preparing to settle on the land and to establish either a 'moshav ovdim' or a 'moshav shitufi'. During their period of training the members of such a group support themselves by working as labourers.

(6) The *'Kvutza'* is based upon the principles of absolute equality between all of its members, and collective production and consumption. The farm is completely independent, the means of production being jointly owned by its members. The dimensions of the settlement are governed by the agricultural potential.

(7) The *'Kibbutz'* differs from the 'kvutza' in that it aims to include industrial enterprises in the economic functions of the village in order to achieve the largest possible population.

Up to quite recently the 'kibbutzim' and 'kvutzot' were organized in two separate movements. Over the past few years, however, the differences between the two have become less clear, as even the 'kvutzot' have expanded beyond the limits their founders considered optimal, mainly as a result of the rapid intensification of farming in Israel.

Today the terms 'kibbutz' and 'kvutza' are virtually synonymous.

(8) The *'Kibbutz' in the 'Moshava'* is a co-operative group, preparing to settle on the land on a collective basis and subsisting, during the period of training, upon the wage labour of its members.

Following the tremendous influx of new immigrants, which was an outstanding feature of the early years of the State of Israel, new forms of agricultural organization appeared, such as the 'Moshav Olim', the 'Kfar Avodah' and the 'Maabara', all of which have been discussed elsewhere in this volume.

The above classification, based upon differences in way of life and in economic concepts, does not cover the entire gamut of distinctions. The shades of difference, distinguishing settlements of the collectivist—and the individualist—type are legion, depending not only upon the optimal dimensions of the collective unit, to which reference has already been made, but upon divergent economic and social ideas, which in turn shape the organizational structure. The collective settlements may be classified broadly as follows:

(1) Those which seek to develop more intimate and closely integrated groups of members. Until some years ago such settlements were termed 'small' or 'organic' 'kvutzot' to distinguish them from the 'large kvutzot'. Settlements of this type were organized in two national federations, 'Hakibbutz Ha'artzi Hashomer Hazair'[2] and the 'Chever Hakvutzot'.[3] More recently the latter has been expanded

---

[1] Organization for settlement (in Hebrew).
[2] The National Collective of the Young Guard (in Hebrew).
[3] Union of Groups (in Hebrew).

and reorganized to form a larger federation, which will be more fully discussed later.

(2) Groups which seek to establish a large and ramified economy, combining agriculture with industry and handicrafts, with a view to absorbing as many newcomers as possible. These settlements are affiliated to the 'Hakibbutz Hameuchad'[1] collective federation.

These federations are further distinguished by their political and social opinions. In the past few years there have been many cases in which political differences within settlements have become so acute as to make withdrawal of a section inevitable. It has become almost normal for a minority dissident group to leave the settlement. The minorities again have combined with others, subscribing to similar political views, to establish new settlements.

The members of the 'Hakibbutz Ha'artzi Hashomer Hazair' have a clearly defined Marxist ideology emphasizing the class struggle, and the national and international unity of all members of the working class. In international politics 'Hakibbutz Ha'artzi' overtly supports the Eastern Bloc. Yet despite the powerful homogeneity of political opinion, under the slogan of 'democratic centralism', its members did not, for many years, form a distinct political party. It was only after the rift in the Palestine Labour Party in 1945, that they united with the dissident group and with smaller, mainly urban, groups of workers to constitute the 'Mifleget Hapoalim Hameuchedet' (United Workers' Party), in which the 'Hashomer Hazair' wing predominated.

The 'Hakibbutz Hameuchad' federation of settlements, based upon socialist principles and national self-determination, was affiliated to the Palestine Labour Party. In the early forties the federation began to organize a separate faction within the Labour Party, leading ultimately to the rift already referred to. Within the 'Hakibbutz Hameuchad' there were settlements, most or all of whose members remained loyal to the Palestine Labour Party. There were others where there was a similar degree of support for the United Workers' Party.

Wherever the two factions were more or less equal in strength communal living proved impossible and it was found necessary to divide the settlements. Five 'kibbutzim' underwent such an operation, new settlements being established for the dissident minority. In a number of cases exchanges of population were arranged, a Labour Party minority in one 'kibbutz' being exchanged for a United Workers' minority in another.

As a result, only 'kibbutzim' comprising a majority of United

---

[1] The United Collective (in Hebrew).

Workers' Party supporters remained within the 'Hakibbutz Hameuc-had'. Those in which a majority supported the Israel Labour Party united with the 'kvutzot' of the 'Chever Hakvutzot' to form the largest of the collective federations, the 'Ichud Hakvutzot Veha-kibbutzim'.[1]

The ideology of the 'Chever Hakvutzot' stressed the ethical aspects of collective living and in this respect resembled socialist movements of an educational and socio-ethical character, in a number of other countries.

At the time of writing dissension in the United Workers' Party has resulted in the secession of a large body of members. The Party is now highly homogeneous in character, the overwhelming majority of its members belonging to the settlements of 'Hashomer Hazair'. The dissident minority, mainly members of the settlements of 'Hakib-butz Hameuchad', has formed a new party called 'Le'achdut Ha'avoda'.[2]

These political changes have at times been very painful for the agricultural settlements. They have been accompanied by exchanges of population, the division of existing settlements and the establish-ment of new settlements. In the majority of cases instead of a single divided settlement, two villages have been established with a double number of settlers. Only with the passage of time can the wounds caused by this controversy heal completely.

The network of workers' settlements also includes other groups with different ideological points of view. These include: the 'Brit Hakibbutzim shel Ha'oved Hazioni';[3] and two religious groups, 'Hakibbutz Hadati shel Hapoel Hamizrachi'[4] and the collectives of the 'Poalei Agudat Israel'.[5] Three 'kibbutzim' are not affiliated to any national federation.

At the beginning of the 1953/54 agricultural year the 216 collective settlements in Israel had a total population of 71,569. These settle-ments can be classified on the basis of period of establishment as follows:

Settlements founded prior to 1935	43
„ „ 1936–40	38
„ „ 1941–47	58
„ „ 1948–53	77
Total	216

---

[1] Federation of the two forms of the collectives: 'kvutza' and 'kibbutz'.
[2] Towards Unity of Labour (in Hebrew).
[3] Federation of Collectives of the Zionist Worker. This is a non-socialist group.
[4] Federation of Collectives of Religious Workers.
[5] Federation of Collectives of the Ultra-orthodox.

The following table shows the number of settlements affiliated to each of the national federations:

'Ichud Hakvutzot Vehakibbutzim'	70
'Hakibbutz Hameuchad'	54
'Hakibbutz Ha'artzi Hashomer Hazair'	68
'Brit Hakibbutzim shel Ha'oved Hazioni'	7
'Hakibbutz Hadati shel Hapoel Hamizrachi'	11
'Kibbutzim of Poalei Agudat Israel'	3
Not affiliated	3
Total	216

Political changes within the labour movement in Israel have not affected the 'moshavim' (or individualist type of settlement) in any way. Most of the members of these villages are affiliated to the Israel Labour Party, and are organized within the 'Tnuat Hamoshavim',[1] which in 1953/54 covered 73% of the settlers, and 76% of the population in the 'moshavim'.

The 'Ha'oved Hazioni', 'Hapoel Hamizrachi' and 'Poalei Agudat Israel' all have independent federations of 'moshvei ovdim'.

The affiliations of 250 'moshavim' in Israel are reflected in the following table (in 1953/54):

TABLE 84

Movement affiliation of 'Moshavim'

	Total no. of settlements	'Moshavim'	'Moshavim Shitufiim'	'Moshvei Olim'
'Tnuat Hamoshavim'	183	66	9	108
'Ha'oved Hazioni'	13	3	3	7
'Hapoel Hamizrachi'	44	10	2	32
'Poalei Agudat Israel'	6	1	1	4
Unaffiliated	4	—	—	4
Total	250	80	15	155

At the beginning of the year 1953/54 the total population of the 'moshavim' was 74,588.

The following table indicates the periods in which the 'moshavim' were founded.

The preference of the new immigrants for the 'moshav' type of village has been striking, and indeed most of those who have decided to engage in farming have opted for the individualist 'moshav olim'.

---

[1] Federation of Co-operative Villages.

TABLE 85

Groups of 'Moshavim' in periods of foundation

	'Moshavim'	'Moshavim Shitufiim'	'Moshvei Olim'
Founded prior to 1935	38	—	—
„ 1936–47	18	4	—
„ 1948–53	24	11	155
Total	80	15	155

THE 'MOSHAV OVDIM'

The 'moshav ovdim' is based upon a number of principles—national land, self-labour, mutual aid and co-operative purchase and sale—which are shared by all types of workers' settlements.

For the members of the 'moshav', agriculture is more than an occupation, it is a way of life. Indeed with the exception of one village, which combines marine fishing with farming, all engage exclusively in agriculture.

A basic principle of the 'moshav' is the freedom of the farmer and his family to engage in whatever branch of agricultural production they choose, implying independent management of the farm and individual responsibility. The planning of the farm, the composition of the various branches of farming, are dependent upon the working capacity of the family unit. Despite differences, sometimes of fairly substantial proportions, in the development of the farms, deriving from the labour potential and skill of the family, this disparity does not normally affect standards of living as expressed in food, clothing, etc. This is largely the result of observance of the principle of self-labour.

In the 'moshav' the status of any settler is not related in any way to his political views. There have been cases of the key position of 'Merakez'—executive-secretary—of the village being placed in the hands of a settler belonging to an opposition political party. There have also been cases of settlers deciding not to be active in the public life of the village, because of political or other differences. The farm, however, provides an adequate and satisfying outlet for individual energies, while the settler can always return to public life when he chooses. The farmer sells his surplus produce through the co-operative society, the proceeds being placed to his credit. But the size of this balance is never permitted to influence the supply of essential services, such as education, health, security, etc.

The community also provides credit facilities to finance farming operations and sometimes even other current needs. In a number of 'moshavim' this aspect of mutual aid is not satisfactorily developed, but even here communal concern for the welfare of the individual prevails.

More recently 'moshavim' are beginning to provide rest and vacation facilities for their member-settlers. For this purpose special funds, in which participation is obligatory, are being established.

The public life of the 'moshav' is designed to secure maximum participation of the settlers in the diverse communal responsibilities.

The supreme public body is the annual general meeting, which elects the Village Council. The council in turn appoints the following bodies: the Village Committee, the Economic Institution Committee (which manages the co-operative store), the Mutual Aid Committee, the Education Committee, the Cultural Committee, the Agricultural Committee (which appoints sub-committees for various branches of farming such as dairying, poultry, hay, agricultural machinery and the like), the Judicial Committee, the Health Committee, the Village Improvement Committee, a Committee to maintain contact with young people of the Village serving in the Israel Defence Army, and Committees for Auditing and for Membership. The total membership of these committees approximates one hundred. It is obvious that every settler who so desires can find a satisfying field of public activity to engage in.

Prior to the calling of the annual general meeting a Nominations Committee negotiates with candidates for the various public offices and prepares proposals for the composition of the villages' committees. These are submitted to the general meeting for approval. The Nominations Committee's task is no easy one, for generally speaking few settlers are interested in holding public office. The first ten to fifteen years after the foundation of the 'moshav' is a period of unremitting exertion. In fact it is only when the children have grown and are capable of contributing to the family labour pool that the parents are able to devote any time to public concerns at all.

THE CONSTITUTION OF THE 'MOSHAV'

Each of the settlement movements has endeavoured to formulate a constitution expressing its specific character and dealing with its peculiar problems. A social movement, it need hardly be stressed, is entitled to make more rigorous demands upon its members than is the civil law.

The constitution of the 'Tnuat Hamoshavim', accordingly, embodies its particular concept of individual rights and duties, including such tenets of the 'moshav' as self-labour and mutual aid. Mutual aid at present is confined to assistance in time of illness or other distress and does not cover such contingencies as an inadequate family labour force, old age or chronic weakness. On the other hand assistance in the provision of credit and the development of the farm is a noteworthy feature.

The 'Tnuat Hamoshavim' is organized upon democratic principles, leaving adequate latitude for discussion and decision.

The following extracts from the constitution of the 'Tnuat Hamoshavim' reflect the principles governing vital aspects of communal life in these villages.[1]

Principles

'Farming constitutes the basis of the "moshav ovdim". The land is nationally owned. The way of life of the members is based upon self-labour, mutual aid and co-operative purchase and sale through the agency of the proper institutions of the "Chevrat Ovdim"[2] and "Nir" [3] (Paragraph 3).

'The "moshav ovdim" is an organized society with a common aim and goal. It forms a nucleus of village self-government and within the area under its jurisdiction it constitutes the source of employment and subsistence for its members. Within this area the community accepts responsibility for the individual. The community must assist in the proper development of the individual farms, must provide for the basic needs of the families of the settlers and organize the public services which every community requires' (Paragraph 5).

Membership

Any 'moshav' resolving at its general meeting to become affiliated to the 'Tnuat Hamoshavim', constitutes in entirety a branch of the Movement. All members of the 'moshav' have complete freedom of conscience and ideology. Membership of the 'moshav' in the 'Tnuat

---

[1] Constitution of the 'Moshavim' Movement, ratified by the Second Session of the Eighth Conference, attended in Kfar Vitkin (name of a large 'moshav' in the Hefer Valley), 2–4 May 1951.

[2] Society of Workers (in Hebrew)—the Central Co-operative Society of the General Federation of Labour in Israel. It is the supreme institution of all economic bodies and companies affiliated to the Federation ('Histadrut').

[3] Furrow (in Hebrew): The Central Society for all agricultural societies affiliated to the Agricultural Workers' Organization. In its authorization and functions it is identical with 'Chevrat Ovdim'. The Marketing Societies are affiliated to both Central Mother Societies: 'Nir' and 'Chevrat Ovdim'.

Hamoshavim' is restricted to the organizational and social functions of the Movement.

'Membership in the "Tnuat Hamoshavim" is collective, thus each "moshav" and all its members become affiliated to the Movement as a single unit.'

'A "Moshav Ovdim" is accepted for membership in the Movement upon application being duly submitted after a decision supported by no less than two-thirds of the number of those participating has been taken at the general meeting of the "moshav".'

Jurisdiction

The General Federation of Labour in Israel maintains courts with jurisdiction in disputes between its constituent institutions and members, in so far as the parties to these disputes consent to the jurisdiction of these courts. The latter operate as institutions of compulsory arbitration. Following the establishment of the State of Israel the jurisdiction of the 'Histadrut' has been restricted to disputes affecting membership in the 'Histadrut'. In all other matters recourse is had to the civil courts.

The customary procedure is for the Village Court to deal with all local disputes. This court is a kind of court of first instance. In more complicated matters, for example, the evaluation of a farm, in case of the impending departure of one of the settlers, or a domestic dispute, the matter is submitted to the Central Judicial Committee of the 'Tnuat Hamoshavim', sitting as a court of second instance. Parties can appeal to the General Histadrut Commission.

'Any dispute or difference between two "moshvei ovdim" (plural), or between institutions of a "moshav ovdim", and any appeal against any decision handed down by a Members' Court in any "moshav ovdim", which will sit as a Members' Court of second instance in keeping with the provisions of the Members' Court of the General Federation of Jewish Labour in Palestine' (Paragraph 1).

'Appeal against any decision or verdict of the Central Judicial Commission, insofar as the right to appeal is recognized by the regulations of the Members' Courts, will be heard by the Supreme Judicial Commission' (Paragraph 2).

Rights and duties of settler-members

The leasehold contract between the national land institution and the individual settler does not constitute an adequate safeguard against deviations from the principles of the 'moshav', nor even for that matter against speculation in land. This, at least, has been proved in a number of housing projects constructed on land owned by the Jewish National Fund. There have even been cases of the sale of farm land at inflated prices for housing purposes. For this reason the 'Tnuat Hamoshavim' has seen fit to formulate certain safeguards to ensure the proper observance of the terms of the leasehold.

'The right of the members of a "moshav" to their holding is conditional upon constant cultivation by self-labour and not hired labour. Under special circumstances the "moshav" Committee can permit the member to cultivate his farm permanently with hired labour, provided the prior approval of the Israel "Moshavim" Movement has been obtained' (Chapter 6, 1).

This paragraph obliges the individual settler to cultivate his land continually with his own labour. In itself, however, it provides no sanction in the event of refusal to carry out its provisions. For many years the 'Tnuat Hamoshavim' has demanded that the 'moshav' as such be the second party to the leasehold contract with the national land authority, and that the settler sign a subsidiary agreement with the 'moshav'. The difficulty in such a case, of course, is that thereby the 'moshav' might become liable for the debts and obligations of the settler. In fact, banking institutions normally grant credits to the village and not to the settlers. The constitution adopted by the'Tnuat Hamoshavim' in 1951 lays down the following provisions:

'The "moshav" as such is the principal leaseholder of the Jewish National Fund or any public authority sub-leasing the farm holdings to its members' (Chapter 6, 2).

In order to provide an additional safeguard against any contingent error or deviation from these principles on the part of the 'moshav', and to guarantee the rights of the settlers the 'Nir Shitufi' Corporation, the body representing the Agricultural Workers' Organization, is a signatory to the contract as a third party.

' "Nir Shitufi" through the agency of the Secretariat of the "Tnuat Hamoshavim" in Israel must also be a signatory to the main leasehold contract between the "moshav" and the Jewish National Fund and any other body as well as to any subsidiary contract that may be signed between the "moshav" and its members, except in such cases as "Nir Shitufi" may decide to appear directly as a third party in such contracts' (Paragraph 2).

The 'Moshav' Movement is interested in preserving the size of

C.P.A.—P   213

the holding in keeping with the family labour force and accepted methods of cultivation in the 'moshav'.

Should methods of cultivation be changed and intensification permit a decrease in the area of the holding, the village retains the right to introduce such changes as it thinks fit. Indeed in many villages such a redistribution of lands has already been effected once, and even in certain cases, twice.

'In the event of agricultural development permitting a working family to subsist on a holding smaller than that originally allocated in any "moshav", the holding must be decreased and additional families settled on the area thus rendered vacant . . .' (Paragraph 4).

This clause is repeated in the internal regulations of the villages, in the following terms:

'The society may, upon the decision of the general meeting, arrange a redistribution of land. In the event of such a redistribution any member who can prove that he has improved his land will receive suitable compensation . . .' (A Constitution for the 'Moshav Ovdim',[1] Part 5, paragraph (a)).

The manner in which the individual settler manages his farm and conducts himself generally is governed by a series of regulations and contracts and particularly by the Tnuat Hamoshavim Constitution, the Regulations of the 'Moshav Ovdim', the leasehold contract with the Jewish National Fund and the contract with the Settlement authority (the 'Keren Hayesod')[2] covering the grant of the initial credits for the establishment of the farm and the purchase of farm equipment.

FATHERS AND SONS

The productive capacity of the farm grows with the increase in the size of the farmer's family. New problems, however, arise when sons and daughters marry. In the older 'moshavim' it is common for a young married couple to live together with their parents, and there are even cases, where methods of cultivation are highly intensive, of two families of the second generation living on the parental farm. One of these families will in the ordinary course of events eventually inherit the holding. The various duties on the farm are allocated by

[1] Approved by the Eighth Conference of the 'Moshavim' Movement, May 1951.
[2] Foundation Fund (in Hebrew)—the financial instrument of the Jewish Agency.

internal agreement, and cases where the intervention of village or public institutions has been necessary, have fortunately been very rare. Under certain circumstances the father of the family may prefer to divest himself of some of the responsibilities of the farm and devote himself to some aspect of communal work. The village welcomes such a decision as it enables it to minimize the extent of hired labour.

In other cases sons or daughers for whom the parental farm cannot provide suitable employment have preferred to settle elsewhere either in a 'moshav' or in a 'kibbutz' or to take up some non-agricultural occupation. An interesting phenomenon in recent years has been the return to the parents' farm of sons and daughters after an interval of many years' duration. In 'moshavim' with reserve lands the second generation have developed their own farms, aided by equipment given to them by their parents.

It is significant that thirty years after the foundation of the first 'moshavim', the second generation are very closely attached to their native villages and no problem exists of sons forsaking the occupation of their fathers. This aspect of 'moshav' life is dealt with in the constitution and especially in three paragraphs cited below which cover the cases of adult children living with their parents, settlement of children within the village and inheritance, respectively.

'Sons and daughters of the village enjoy priority in settlement on the "moshav lands" ' (Chapter 6, Paragraph 5).

'All adult members of the families of settlers (above the age of eighteen, including dependants by marriage) who live and work in the village, and engage in agriculture or any other occupation, enjoy full voting rights to all institutions of the "moshav" ' (Chapter 6, Paragraph 7).

'In order to ensure the continuity of the family holding from generation to generation, the Movement regards the "moshav" farm as the home of at least two families—of the parents and one of the children together. In determining the unit of land per family this contingency must be taken into account, and in the event of the farm not being capable of supporting two families other sources of livelihood must be found. The "moshav" Committee must assist in finding these sources' (Chapter 6, Paragraph 6).

INTEGRITY OF THE FARM

The individual holding is a prime consideration of the 'moshav' movement in dealing with problems of family and personal status and it seeks by various means to ensure the integrity of the farm.

In the case of separation or widowhood the 'moshav' uses its good offices to ensure the rights of the party remaining in the village. In the case of a person no longer capable of managing his farm the Village Council may decide to transfer the farm to a new settler, conditional upon due provision being made for the departing member.

INHERITANCE

In keeping with the State inheritance laws—themselves a legacy from other political regimes—the farm of a deceased person, like all other assets in his estate, must be divided among the surviving heirs, whether the latter live on the farm or not (they may even be resident abroad). In the event of the matter being submitted to the courts of law the farm may be so sub-divided that it ceases to be a viable unit.

Provision has been made in the leasehold contract to meet such a contingency by approving mutual agreement between the heirs in order to maintain the integrity of the holding.

'Members of the "moshav", or their children may not divide the farm among themselves.'[1]

The leasehold contract becomes null and void following the death of one of the lessees, in the case of divorce or informal separation, or dissension between the lessees or their families which prevent the proper working of the farm.[2]

Up to the present no cases have been reported of any clash between the letter of the inheritance law and its practical application in the 'moshavim'. However, such a possibility must be foreseen. The custom is that the family itself is entitled to decide who shall take over the undivided holding. In the absence of mutual agreement, however, the decision rests with the village.

THE 'MOSHAV OLIM'

The aspects of the constitution of the 'Tnuat Hamoshavim', the 'moshav' regulations and the leasehold contract, discussed in the foregoing, apply to the 'moshav ovdim'. The regulations of the 'moshav olim', in addition to the provisions which are common to all types of settlement, possess a number of special features.

It must be stressed that the 'moshav olim' differs from the older 'moshavim', which were established by settlers who had undergone a lengthy period of preparatory training in an 'Irgun',[3] or by native-born Israelis. In the initial years the settlers of the 'moshav olim' support themselves mainly by outside labour. Most of them have

[1] Leasehold Contract, Clause 11, item a.
[2] Ibid., Clause 12, cl. 1, 5, 6; cl. 2, 3.
[3] Preliminary training group.

216

had no agricultural training, and in the majority of cases no previous experience even of manual work. They lack, moreover, social and organizational education and experience. For this reason special regulations govern arrangements in the period in which such a 'moshav' is being developed. Wages are computed as far as possible on a basis of piece-work, according to a scale designed to prevent excessive differentiation in income. The rates for various types of work are based upon a general average. A certain percentage of the piece-work earnings is deducted and paid into an 'Equalization Fund' to raise earnings of those employed upon a daily wage basis. The wages are paid to the village, which in turn pays them out to members. Other clauses govern the administration of the village, conduct at general and committee meetings, attitude to the 'moshav' and the preservation of the family. There is a total prohibition of all games of cards or hazard, drunkenness, breaches of the village peace and the like.[1]

THE WOMAN IN THE 'MOSHAV'

A role of vital importance is reserved for the woman in the 'moshav' and her contribution is often a decisive factor in the success or failure of the family farm. In addition to her multifarious tasks in the home, the kitchen and in care of the children she has many other responsibilities, such as the preparation of mash for cattle and collecting eggs, milking the cows, cultivation of the vegetable garden and especially the harvesting of ripe vegetables.

In the statutes, the regulations and the leasehold contract her status is recognized as equal in every respect to that of her husband and it is expressly stated that wherever the term 'chaver' (member) is used the provision applies equally to the 'chavera' (female member). Paragraph 13 of Chapter 2 of the constitution provides for the 'activization of members, both male and female, in all public activity connected with the "moshav ovdim", the Federation of Labour, the Movement, the State and the Zionist Movement'. Women, indeed, play a prominent part in the life of the village, in local institutions and even to some extent in institutions of a national and general nature. The 'Tnuat Hamoshavim' was represented— within the Labour Party—by a woman in the First and Second 'Knesset'.

Contrary to the clause which lays down that 'mutual aid, in labour or in money, which is extended to farmers (in the form of labour in times of sickness, in money in the event of any accident on the farm) does not include public officials', a clause immediately following

[1] Regulations for 'Moshvei Ovdim', Chapter 10, Clause e.

provides that 'mutual aid in labour or in money among the women of the "moshav", includes all women'.[1]

Very few clauses in the regulations refer specifically to women. In the chapter dealing with mutual aid, different rosters are laid down for men and women respectively. One clause declares that 'In every case of birth the society shall extend aid to the family of the mother by providing hired help for a period of not less than a fortnight.'[2]

In the leasehold contract the reference is to the lessees, specifically including all members of the family who are entitled to be signatories to the document, including the settler, his wife, sons, daughters-in-law, daughters, sons-in-law.

In the event of the lessee-family breaking up the contract is rendered null and void. Mention has already been made of the voiding of the contract in the event of divorce or even informal separation.

The party or parties affected by the voiding of the contract may apply to the Judicial Committee to decide who shall be given priority in the signing of a new leasehold contract. There are not a few cases of farms being managed by women, either divorcees or widows.

COMMUNAL EMPLOYEES

Communal employees were full members of the early 'moshavim'. Teachers, for example, paid their salaries, received from the national education authority, into the communal chest, and were paid according to the locally determined rate. The status of such public employees, however, changed fundamentally as their number increased. In some cases settlers, who had given up farming, took up various positions in the 'moshav', but in the majority of cases vacancies were filled by candidates from outside the village whose salary was fixed without respect to local standards of production and income. Farmers and public officials live according to different standards. In many cases members of the families of these officials do not work. The desire to integrate the employees of the 'moshav' into its economic structure has not been realized, and today there is even less prospect that this can ever be done, mainly because the farmers will never allow public employees who are not themselves engaged in agriculture to participate in the planning and administration of the village economy. No practicable method has so far been evolved of utilizing the average earnings of the settlers as a basis for computing the salaries of these officials and employees, in order to ensure a maximum degree of economic equality. The

[1] Constitution of the 'Tnuat Hamoshavim', Appendix A, Chapter f, Clauses 17, 18.

[2] Ibid., Part 6, Clause 1, Para 2.

218

employees of the 'moshav' are members of their respective trade and professional unions, and the accountants, the teachers, the mechanics, the clerks employed in the co-operative store, all claim remuneration according to rates laid down by these unions. The settlers, accordingly, oppose acceptance of public employees as full members of the 'moshav', and have suggested that their status be that of residents, as long as their duties in the village require it. We are not of the opinion that the public employees of the 'moshav', who live under different conditions from those of the agriculturists, can justly claim any say in determining the life of the 'moshav'. The inherently contradictory nature of this demand must be appreciated, for in effect what these public employees were asking for was full membership in the 'moshav', with all the privileges that go with such membership, and simultaneously membership in their trade or professional union, to protect their interests—including wage rates, vacation, conditions of dismissal and the like—vis-à-vis their employers, namely the 'moshav'.

Negotiations over this issue continued between the parties for a number of years. The question was debated during the 'Moshav' Conference[1] and was submitted for final decision to the Executive of the General Federation of Labour.[2] The latter's findings were in the nature of a compromise. The public employees were recognized as members of the 'moshav' and as such as members of the 'Moshav' Movement. Their remuneration was fixed according to current salary rates in the Federation of Labour. The 'moshavim' were required to establish smallholdings for their public employees but part of the income of the holding was to be deducted from their salaries. All disputes regarding the status or working conditions of public employees must be adjudicated by a committee representing three central institutions, the Executive of the General Federation of Labour, the Agricultural Centre and Central Control Committee.

Six years have passed since these recommendations were made and it is interesting to note that the controversy had since died down.

The new constitution for the 'moshavim' adopted in 1951 includes clauses governing the status of public employees drafted in the spirit of the decision handed down by the Executive of the Federation of Labour. The following extracts from the relevant paragraphs are noteworthy:

'Permanent appointment of a public official in the "moshav ovdim" is conditional upon his being a member of the "moshav".'[3]

[1] Held in Beer-Tuvia, October 1945.   [2] February 1948.
[3] Constitution of the 'Tnuat Hamoshavim', Appendix A, Chapter b, Para. 1.

'Permanent public employees of a "moshav ovdim" whose appointment dates at least three years prior to the approval of this Constitution . . . automatically become members of the "moshav" except where such membership is opposed by one-third of the members of the "moshav".'[1]

'Should any public employee be accepted as a member of a "moshav" the latter must provide him with a plot of land, and must assist him by the grant of a loan and other means in the development of a smallholding . . .'[2]

'Public employees must sign a contract with the "moshav" subleasing their holdings in accordance with the regulations of the "moshav".'[3]

The constitution lays down that the status of public employees shall be equal to that of the members of the "moshav", their "employers", in the following terms:

'Public employees who are members of the "moshav" shall have full voting rights in all meetings of the "moshav", the right to elect and to be elected to any office in the "moshav" or in the "Tnuat Hamoshavim".'[4]

'Responsibility for the adult sons and daughters of members of the "moshav" who are public employees or independent professional workers devolves upon the "Moshav" Committee and they must be dealt with in the same way as the children of settler members who are not absorbed in the "moshav". The "Tnuat Hamoshavim" shall accept responsibility for organizing them for settlement on the land.'[5]

A comparatively large number of clauses deal with the position of public employees, their pensions, compensation rights, mutual aid, payment of rates and investments, transfer from one position to another, as well as their rights in the event of their leaving the 'moshav'. In this case they shall receive all sums to their credit in the provident fund. The provisions of the Constitution are consistently observed by the 'moshav'. The public employees, it seems, regard it more in the nature of collective agreement with their employers and have not exercised the privileges it has granted them.

Not all of them, for example, have developed smallholdings. The 'moshavim', on the other hand, have not availed themselves of the right to deduct part of the income the employees derive from their

---

[1] Constitution of the 'Tnuat Hamoshavim', Appendix A, Chapter a, Para. 4.
[2] Ibid., Chapter b, Para. 7.   [3] Ibid., Chapter 9, Para. 8.
[4] Ibid., Chapter d, Rights, Para. 12.
[5] Ibid., Chapter d, Para. 13.

smallholdings from their salaries. It must also be noted that membership of the employees in the 'moshavim', up to the present at least, has not persuaded them to identify themselves more completely with the village. On the contrary, few of them evince any active interest in the affairs of the villages in which they live and work.

INDEPENDENT PROFESSIONAL WORKERS AND SKILLED ARTISANS

Skilled artisans in the 'moshav', such as the smith, the carpenter, the shoemaker, and at times the truck driver, do not receive a fixed wage but are paid individually by whoever has recourse to their services. They have specified rights but unlike the public employees do not constitute a distinct sector of the community. They are full citizens of the village, have developed smallholdings and are rapidly adapting themselves to 'moshav' life.

NON-MEMBER RESIDENTS

The 'moshav' endeavours to maintain complete identity between co-operative organization and municipal organization, and all decisions taken in regard to rights or duties of members are equally applicable in both spheres.

In the course of time persons employed elsewhere as railwaymen, road workers, foresters, guards, etc., have taken up residence in 'moshavim' because of the proximity of the latter to their place of employment, or because the 'moshav' provides a more suitable environment in which they can bring up their children. This category also includes former public employees who continue to live in the 'moshav'. These residents possess no direct ties with the 'moshav' and for this reason, indeed, have no rights in it. They do, however, pay rates. In the 'moshava' and, of course, in the city, under similar circumstances, they would be granted full municipal rights. In the 'moshav' the situation is more complex because, as already indicated, the municipal aspect in this type of village is inextricably bound up with its agricultural operations. It would be strange, to say the least, to grant any resident a voice in economic affairs purely because he pays the Education or Water Rate. On the other hand, however, it is democratically impossible to deprive such a class of residents in the 'moshav' of their elementary civic rights. Leaders of the 'Moshav' Movement, indeed, are of the opinion that such rights should be recognized but that the residents should also assume certain responsibilities. If they should be taxed upon the same basis and according to the same scale as other members of the 'moshav', the influx of residents who have no special interest in living in the 'moshav' would be checked.

RATES

So far no standardized system of rates has been developed in the 'moshavim'. In 'moshavim' where members' income is derived from outside work the system of progressive taxation is common. Where the members support themselves by farming, two systems, equal taxation and progressive taxation, operate side by side. Though equal taxation has already struck roots in the structure and organization of the 'moshav', however, there is a growing tendency to introduce progressive taxation. Possibly State income tax, which is now being imposed upon the 'moshavim', too, will accelerate the adoption of the progressive system.

LIFE IN THE 'MOSHAV'

The pioneers, the ideologists, the founders of the 'moshav' in Israel are our own contemporaries, and this generation has borne witness to the onerous burden they assumed, particularly in the initial period of colonization. The daily chores of farming took up all of their time, though they themselves never intended engaging exclusively in manual labour and abstaining completely from all interest in things of the spirit, in literature or art.

A clause inserted in the leasehold contract obliges the settlers to observe the Sabbath and the Jewish festivals.

'The lessees must rest on the Sabbath and all Jewish festivals and must not engage, within the boundaries of their farms, in any building work or any work in field or vineyard, in industry or in commerce. . . .'[1]

This clause, of course, does not include all tasks on the farm. For example care must be taken of livestock on the Sabbaths and the festivals. But during the initial period, both before and after the signing of the leasehold contract, when the main source of income was outside employment, the settlers were unable to observe its provisions fully and on the Sabbath performed tasks which they were unable to do during the week.

As the village developed the school began to play a role of increasing importance in the observance of national holidays and festivals. Parents naturally derived much pleasure from the sight of their children participating in shows, choirs, orchestras, dance troupes

[1] Leasehold Contract, Para. 13.

and the like. A tradition of folk festivals developed around such holidays as the Feast of the First Fruits, Israel Independence Day, the common Seder on Passover, Purim balls, as well as celebrations to mark various local and regional anniversaries.

In recent years the custom has developed of inviting outside artists to assist in these festivities and not infrequently the larger theatrical groups visit the country districts. Orchestras, choirs, singers, etc., do likewise. Lecturers are sent regularly by central educational institutions, and particularly by the Hebrew University's People's Education Department. Often the 'moshavim' invite lecturers upon their own initiative. Every 'moshav' maintains a library. The Beth Ha'am[1] is a common institution in the older 'moshavim', but unfortunately owing to the excessively high costs of building other villages are not able to erect similar institutions. Books enjoy a good sale in the villages and the practice of former years for two families to subscribe jointly to a newspaper has given way to individual subscription. Most families also subscribe to weeklies and monthlies of an agricultural or political character.

Short courses in a variety of subjects, including agriculture, dramatics, dancing, choral singing, etc., are organized from time to time and in many 'moshavim' there are groups which meet regularly to discuss books, to study the Bible or art. The winter with its long evenings is the favourite season for these activities.

The younger generation live their own life in the 'moshav'. Most of the youngsters are members of the Noar Oved[2] Organization and participate in its camps, seminars, etc. The 'Tnuat Hamoshavim', too, devotes special attention to the younger people and also organizes various study courses on social and political questions. Some of these courses are based upon an annual two months' period of study over three or four years. The sons and daughters of the 'moshavim' made a conspicuous contribution to the national war effort during the Second World War and the Israeli War of Liberation, and many are now assisting in the absorption of newcomers by serving as instructors in the new immigrants' settlements.

A survey covering five 'moshavim' with 443 families, having 622 sons and 560 daughters, of whom 373 sons and 385 daughters are adults and 174 sons and 251 daughters are already married, was conducted recently to ascertain the attitude of the younger generation toward settlement on the land. The results of this survey are summed up in the following table.

This table proves that 93% of the sons and 72% of the daughters of these five villages were engaged in farming either in

---

[1] A folk house.   [2] Working youth.

TABLE 86

Present place of residence of sons and daughters of farmers
in five 'Moshavim' (%)

	Sons	Daughters
On parents' farms	60·9	32·0
On their own farms in their native village	6·0	5·6
Otherwise employed in their native village	6·0	6·7
Settled in another 'moshav'	8·3	12·6
In 'kibbutzim'	0·8	6·9
In the army	11·0	8·2
Not in villages	7·0	28·0

their parents' homes or elsewhere. It seems that a considerable
number of young women who have married outside their own
'moshavim' no longer work in agriculture.

There have been cases where because of domestic differences the
parents have remained without anybody to take over the farm,
especially after the marriage of their children. No solution has so far
been found to this problem, for the village refrains from intervention
in internal domestic matters.

THE 'MOSHAV SHITUFI'

The 'moshav shitufi' must be regarded as a synthesis of the principles of the 'kibbutz' and the 'moshav' type of settlement, and has
passed through a number of stages of development before crystalliz-
ing into its present form. The distinctive trait of the 'moshav
shitufi' is that production is planned on collective lines—as in the
'kibbutz'—while consumption is individual, as in the 'moshav'. The
farm is jointly owned, all planning, organization and administration
of agricultural—and other—work being done by a central committee
elected by and representing the community. The working population
are allocated to their various tasks by the 'sadran avoda', the work
manager, whose authority also extends to the womenfolk, after due
allowance has been made for the time required by each woman to do
her housekeeping and other domestic tasks. The household is the
domain of the family and is run in keeping with its tastes and prefer-
ences. In some 'moshavim' of this type industry and handicrafts are
part of the village economy. The majority of the founders of the
'moshav shitufi', it is interesting to note, were former members of
'kibbutzim', who sought a new and more satisfying way of life. The
'moshavim shitufiim' are affiliated to the 'Tnuat Hamoshavim', but
there are still differences outstanding between them and the Secretariat of the latter. The main point at issue is the insistence of the

Secretariat that a small farmyard plot, four dunams in extent, be planned around the cottage, to be cultivated either individually or collectively, the object being to allow the members a loophole, should they eventually prefer a 'moshav' type of settlement. A decision to this effect would become operative if supported by two-thirds of the members of the settlement. The members of the 'moshavim shitufiim', however, refuse to allow any room for a retreat from the way of life they have chosen.

There is also one fishing village, developed upon 'moshav' lines, by a group of ex-servicemen. Settlers in this village are divided into a number of crews of vessels, who own their boat, allocate the various tasks among their members by internal arrangement and share the profits equally. This form of organization is unique. Some of the settlers cultivate smallholdings near their cottages.

THE 'KVUTZA'

A considerable literature has already accumulated regarding the various aspects of the 'kvutza'. This includes sociological and economic studies, fiction and poetry, not to speak of thousands of articles published in the press, innumerable speeches printed, internal news-sheets, etc. In spite of this, no single definition, setting forth the essence of the 'kvutza' in a nutshell, has yet been formulated. The present attempt does not pretend to be more than a vocabulary definition, and while probably not adequate appears to suffice for the purposes of the present chapter on the sociology of the Israel village.

A 'kvutza'—or the 'kibbutz'—is a group of individuals who have united voluntarily for the purpose of establishing in Palestine a homeland for the Jewish people based upon socialist principles, and who, to achieve this goal, have created an economic and social, mainly agricultural, unit embodying the principles of complete equality, mutual responsibility, self-labour and the negation of private property and the organization of all production and consumption on a collective basis.

The 'kvutza' requires every member to make his contribution in keeping with his capacity and allows him to receive from the common fund in accordance with his needs. This basic precept applies under all conditions and circumstances, the sole limiting factor being the capacity of the collective. In other words equality is the criterion in mutual relations in this collective partnership of work. (It is the keystone of the structure of the 'kvutza' and is the active element in the other principles like agriculture, intensive and large-scale mixed farming on a collective basis.)

Today the terms 'kibbutz' and 'kvutza' are to all intents and purposes synonymous. Previously, however, they distinguished between two types of collective settlement, differing in dimensions, economic organization (whether the settlement should be based solely upon farming or should include industry and handicrafts) and the degree of authority exercised by the central federations of settlements. The 'kibbutz' was the large settlement with a mixed economy, while the 'kvutza' was based solely on agriculture.

Under the impact of external factors, however, these shades of difference vanished and a far greater degree of homogeneity emerged, if not among the individual 'kibbutzim', then among the 'kibbutz' federations. But as economic distinctions receded into the background, political differences sharpened, leading eventually, as a result of a split in the Israel Labour Movement, to rifts in the national settlement federations, and even within the settlements themselves. A number of 'kibbutzim' broke away from the 'Hakibbutz Hameuchad' as a result and united with the 'Chever Hakvutzot', to found the 'Ichud Hakvutzot Vehakibbutzim'.

It was the creation of this new body that finally destroyed the distinction between the two terms. As a point of fact there are 'kvutzot' which operate factories and 'kibbutzim' which do not.[1]

The fundamental idea at the root of the collective way of life is thousands of years old and derives from religious practice and beliefs. The purpose of monasticism, throughout the ages, has been to enable a group of religious persons to seclude themselves from secular cares and to devote themselves entirely to the service of God. It has been claimed that Vedism and Brahmism were the first religions to adopt the monastic way of life, which was later taken over by Buddhism. Buddha himself was the head of such a 'sanga' or monastic group. The Muslim dervishes borrowed the principles of monasticism from India and Thibet. The monastic way of life, however, was most highly developed in Christendom, which in the Middle Ages developed a ramified system of monasteries. It may well be that Christianity was influenced in this direction by the example of the Essenes, an early Jewish ascetic sect, whose achievements in the organization of collective living were notable. The Essenes, it is interesting to note, did not rest content with organizing themselves into a distinct sect, but based their entire way of life upon collectivist principles with the object of generating a mass collectivist movement.

These examples exerted a powerful influence upon the religious collectivist communities, which sprang up in North America during

---

[1] Among the 'kvutzot' which have factories are Sdeh-Nehmeia (plastics), Hanita (precision mechanics), Mishmarot (plywood), Genigar (raffia products), Mezuba (textiles), etc.

the eighteenth and nineteenth centuries. Communal colonies based upon socialist principles were also established at the time. For the most part these colonies, with the exception of two, which existed for 155 and 120 years respectively, were shortlived. Some of them became very powerful but declined as a result of the indifference of the younger generation to the ideals which had inspired their parents. There were other causes, too, including changing conditions, the economic development of North America and the emergence of the competitive system and extreme forms of individualism.

The influence of these experiments in collective living cannot be ignored in tracing the origins of the 'kvutza' movement.

The 'kvutza' is based upon principles formulated forty-five years ago, when the first collective settlement in Palestine was founded, and which have not changed essentially despite the vicissitudes the country has experienced in this period.

The conditions and motives which led to the creation of the 'kvutzot' in Palestine differed fundamentally from those which led to the creation of the religious and socialist communal colonies in other countries. It is a significant fact that their foundation was not heralded by any definite ideology and that to this day the immediate author of the collectivist form of settlement in Israel is not known. Projects for the launching of a collective farm colony, which was to be owned by the Jewish community of Jerusalem, were submitted to Sir Moses Montefiore, upon his second visit to this country in 1839. The plan was not put into effect, and other such schemes, formulated in 1868, 1880, 1881 and 1883, shared a like fate. Quite possibly similar projects were mooted at an even earlier date, though no record of them remains.

The main obstacle to the realization of these plans was the system of 'Haluka'[1] then current in Palestine and particularly in Jerusalem. The Old Yishuv had evolved a parasitic form of existence based upon the collection of alms from among the Jews of the Diaspora for the support of synagogues and 'yeshivot'—schools for the study of Jewish religious lore—in the Holy Land. The almoners of the Jewish population were bitterly opposed to any attempt at constructive action, and it was undoubtedly a result of their efforts and influence that plans for the foundation of villages or other productive undertakings came to nought. They feared that the funds coming from abroad might be diverted to these projects to the detriment of the 'Haluka'.

The regulations governing the establishment of the early Jewish colonies in this country in the eighties of the last century incorporated

---

[1] Charitable contributions.

certain elements of collectivism, but no attempt was made to imple- ment them. The regulations adopted by the Bilu settlers speak of joint ownership of the soil, of the prohibition of private property, the establishment of a common fund and store, etc. It is an historic fact that the members of this group lived communally while they were employed as labourers. When they founded their village, how- ever, they did nothing to implement the clauses envisaging collective organization. Lack of reinforcements was the immediate cause of the failure of the first attempt to settle on the land, as it was of the early uncertain experiments in collective living.

Political Zionism, which towards the close of the nineteenth century succeeded the Chibat Zion[1] Movement, generated a new enthusiasm and a more vigorous approach to the problems of colonization in Palestine. The general atmosphere was now far more congenial for experiments in collective settlement on the land. It seems that conditions reigning in this country fifty years ago were favourable for the development of the 'kvutza', and indeed it has been stated that only colonization on collective lines could have suc- ceeded in laying the foundations of the new work of Jewish settle- ment.

The roots of the Zionist ideal go far back into the remote past and may originate in the Babylonian exile, about 2,500 years ago. Throughout the ages, however, it has assumed diverse forms in keeping with the political and cultural conditions under which the Jews lived. In the Middle Ages it took on the character of mystic messianic movements, envisaging a miraculous return to Zion. In the modern era democratic movements of national liberation in nineteenth-century Europe exerted a powerful influence upon Zionist ideals. The idea of national auto-emancipation was developed by Jewish intellectuals who had absorbed the liberal philosophy of Western Europe. Some of these intellectuals sought to combine Zionist ideals with the desire to return to manual labour and pri- marily to tillage of the soil. The difficulties attending such a process were almost insuperable. Young Jews, totally unaccustomed to physical work, found themselves in a strange climate and environ- ment; and without any funds of their own. But these difficulties only served as stimuli for the development of collectivism. Their only prospect of surmounting the obstacles which confronted them was by a concerted effort.

The land institution of the Zionist Movement, the Jewish National Fund, was founded in the early years of the present century. The tracts which the Fund succeeded in acquiring were, for the most part,

[1] Love of Zion.

228

fallow and had suffered as a result of generations of neglect. Before they could be brought under the plough they had to be improved and drained. The most effective means for securing this object was to transfer the land to a band of these young enthusiasts who indeed embarked upon their task with inspired vigour. At first these groups counted no more than eight to ten members and constituted a substitute for the family environment which the young pioneers had left behind. Gradually they developed an attachment for their place of work, while simultaneously the bonds between the members grew stronger.

At the close of the First World War optimism regarding the prospects of extending the work of settlement on the land was widespread. The cessation of hostilities in Europe was followed by a large influx into Palestine of idealistic pioneers, who constituted natural reinforcements for the collectivists. The new tasks of colonization called for new instruments. It was to meet this need that the large collective settlement, the 'kibbutz', was conceived. Under the impact of conditions that had obtained during the war, the idea of labour service for the regeneration of the Jewish people in Palestine was developed, and the 'Gdud Ha'avoda'—the Labour Brigade—to engage in all sorts of work, building, public works, agriculture and the like, was organized. The 'Gdud', which was short-lived, was composed of 'Plugot'—companies, operating in various parts of the country. Following the inclusion of settlement among its functions, the 'Gdud' disintegrated into independent groups. The large 'kibbutz', accordingly, was the lineal descendant of the 'Gdud Ha'avoda'. In the course of time the collective 'kvutzot', which for ideological reasons had insisted upon a more restricted membership, grew beyond the limits foreseen by their founders, and today, as already stated elsewhere, there is no longer any essential difference between the terms 'kvutza' and 'kibbutz'.

Table 87 on page 230 gives the figures for the number of collective settlements and their population reflects the growth of the collective movement.

Table 88 on page 230 shows the growth in population in the five collective settlements founded prior to the outbreak of the First World War. The sixth, though included in the list of 'kvutzot', was from the very outset rather a co-operative farm. Later it adopted the 'moshav ovdim' form of organization, counting ten families, who constituted the nucleus round which a medium-sized 'moshav'—Karkur—developed.

Gan Shmuel, Merchavia and Hulda all suffered intermittent crises, as a result of which the early settlers left and were replaced by newcomers.

TABLE 87

Growth of 'Kibbutzim' and population

	No. of settlements	Population
Founded prior to 1914	6	179
,, ,, ,, 1927	26	2,277 (341 persons in the 6 settlements founded before 1914)
,, ,, ,, 1938	59	14,223 (7,280 persons in the 26 settlements founded prior to 1927. Two settlements ceased to exist as 'kvutzot')
,, ,, ,, 1953	216	71,569 (35,798 persons in the 59 settlements founded prior to 1938)

TABLE 88

Growth of population in five oldest 'Kibbutzim' during 40 years

'Kvutza'	Population 1914	Population 1953	Index (1914 = 100)
Kinneret	49	749	1,528
Dganiya A.	30	402	1,340
Gan Shmuel	30	586	1,953
Merchavia	20	580	2,900
Hulda	50	327	605

Hulda was razed during the 1929 disturbances and was reconstructed by a new group. Merchavia was the site, shortly after its foundation in 1911, of an agricultural experiment known as the 'Co-operative of Merchavia'. Neither Kinneret nor Dganiya, however, experienced changes of population of this kind and many of the surviving founders are still resident in these settlements. This, of course, is not meant to imply that many members of both did not leave owing to their inability to adapt themselves to collective conditions.

The 'kvutza' has also served as a sort of transit station for new immigrants, who later settled elsewhere. Many of these newcomers never seriously entertained the prospect of remaining in the 'kvutza' for any length of time. Others were compelled to leave because of extraneous reasons, such as the inability of the 'kvutzot' at that time to extend adequate aid to the parents of their members. Today this reason is no longer operative and the 'kvutzot' are able to make satisfactory provision for the parents of their members, though the parents themselves do not always find the collective environment congenial. Another reason was the inability of single members to find husbands or wives. Political differences have also been a cause of members leaving the 'kvutza'. Upon the twentieth anniversary of the first two 'kvutzot' in the country it transpired that 42% and 30% respectively of those who had been members in the first decade had left. But it must be borne in mind that both of these settlements had served as a training ground for other 'kvutzot' and even for other movements. The founders of the first 'moshav ovdim' and of the 'Tnuat Hamoshavim' were former members of Dganiya, while the ideologists and founders of the large 'kibbutz' came from Kinneret. During this period the General Federation of Jewish Labour in Palestine was founded. Among its first leaders and secretaries and directors of its various institutions were many members of Kinneret whose reason for leaving their 'kvutza' was not fatigue or frustration but rather an ambition to launch other pioneering projects often on a far larger scale.

The pioneers of the collective movement were inspired by a desire to change the social structure of the Jewish people. They felt that this could be brought about only by territorial concentration in Palestine, a return to manual labour and the creation of conditions for national independence. The establishment of a network of agricultural settlements was to serve as an instrument towards this end. It is highly significant that forty years after the foundation of the first 'kvutzot', a number of the original pioneers were among the leaders and legislators of a new Government of Israel.

From the very outset the 'kvutza' constituted an integral organ of the new Jewish community in Palestine. This, indeed, is a major point of difference between the Jewish 'kvutza' and religious and socialist communal colonies founded in other countries. The 'kvutza' was not developed solely as an instrument for realizing the principles of socialism and social equality. Jewish socialists in Palestine could have sought to realize their ideals in this country by methods similar to those adopted by socialist movements elsewhere. The prime motive of the 'kvutza' was to accelerate the Jewish renaissance, to renew Jewish nationhood. For this reason its members preached and practised a return to manual labour and cultivation of the land; they aimed at the re-education of the people and the acceptance of responsibility for the community by the individual.

The exigencies of a lifelong partnership with other members in the 'kibbutz' and the consciousness of responsibility for the nation's future, mean that the individual takes upon himself tasks and duties, within a certain framework, which he can only perform as a result of voluntary decision and free choice.

During a later stage, when the membership of the 'kvutzot' began to grow, it was found necessary to commit the rights and the duties of the members to carefully formulated statutes. The creation of national federations was a natural outcome of the increase in the number of collective settlements and the divergent political and social ideologies which began to distinguish them.

The undue stress placed upon political differences in the collective has been and still is the subject of severe criticism. At the same time it must not be forgotten that the settlement federations have made an invaluable contribution in organizing and uniting the 'kvutzot', in guiding them in the agricultural and social spheres. There have been cases where timely assistance extended by a settlement federation has enabled individual 'kvutzot' to weather a passing storm and has even preserved them from disintegration.

An important function has been fulfilled by these federations in improving the organizational structure of their constituent 'kibbutzim' and strengthening them economically. The complaint that they encroach upon the authority of the Agricultural Workers' Organization is unfounded. It is true that their existence does give the latter a federative character, but it is equally obvious that the Agricultural Workers' Organization is incapable of providing the intimate bond essential for the collective movement. The diverse instruments which these federations have established for work in various fields—economic, social, organizational and technical—have proved highly efficacious and have provided important assistance to the veteran and the young 'kvutzot' alike.

The labour movement rightly strives for complete unity, but such unity does not exist. The split in international labour has left its mark everywhere, and is largely the cause of the split in the Israel labour movement. It is pointless, therefore, to complain about the introduction of political differences into the 'kibbutz' movement. We have simply to accept this as a fact, and to try, so far as possible, to derive what advantages we can from this unfortunate situation. It is to the advantage of the 'kibbutz' to be associated with a wider public. Thus, the 'Hashomer Hazair' Federation of 'kibbutzim' is connected with the United Workers' Party—'Mapam'; the 'kibbutzim' of the 'Hakibbutz Hameuchad' Federation are mainly associated with the new party 'Le'achdut Ha'avoda'; and the 'kibbutzim' of 'Ichud Hakvutzot Vehakibbutzim', as well as the 'Moshav' Movement, are associated with the largest workers' party in Israel, 'Mapai', which is also the largest party in the Government coalition. All these federations of villages belong to the Agricultural Workers' Organization and to the General Federation of Labour of which it forms a part. Thus the 'kibbutz' or 'kvutza' receives support and assistance through its association with all these various organizations.

Regulations and statutes

In 1925 a Conference of 'kvutzot' resolved to form the 'Chever Hakvutzot Vehakibbutzim', affiliated to the General Federation of Labour. The object of this new organization was to foster the growth and the reinforcement of the collective movement in town and country, by increasing its absorptive capacity and training newcomers for collective living. It was also to establish relations between the constituent 'kvutzot' upon the collective principles of mutual aid and responsibility and an equal standard of living. To achieve these objects suitable members of the existing 'kvutzot' were to be seconded to positions of leadership in the expanding 'kvutza' movement. The Conference, be it noted, represented all sections of the collective movement and the federation it created was not only representative but fully comprehensive. But two other 'kibbutz' federations were in the process of formation at the time, both of which later resolved to secede from the 'Chever Hakvutzot Vehakibbutzim', one to found the 'Hakibbutz Ha'artzi Hashomer Hazair' and the other 'Hakibbutz Hameuchad'. The remaining 'kvutzot', organized within the 'Chever Hakvutzot', were reinforced, as already related elsewhere, by a number of settlements which decided to secede from the 'Hakibbutz Hameuchad' in 1951, the united body taking the name of the 'Ichud Hakvutzot Vehakibbutzim'.

The regulations and statutes of the collective movement are formulated in three documents:

(1) The regulations of the 'Ichud Hakvutzot Vehakibbutzim' which is at present in the form of a Programme for Unification. This Programme has superseded the regulations of the 'Chever Hakvutzot', but has not yet been incorporated in a written Constitution.

(2) The regulations of 'Hakibbutz Hameuchad', adopted in 1927, which will probably be revised following the split in this federation.

(3) The regulations of 'Hakibbutz Ha'artzi Hashomer Hazair', adopted in 1933, which still remain in force.

The structure of these documents varies and though most of their principles are identical in spirit the formulation is different.

A 'Kvutza' Constitution, for all 'kvutzot' within the framework of the General Federation of Labour, has been drafted. The basic principles, according to Statutes of the 'Histadrut', are as follows:[1]

(1) Negation of private ownership of the land. The land must be owned by the nation.

(2) Self-labour (the 'kvutza') and the personal obligation of all members of the 'kvutza' to engage in work.

[1] *Statutes of the Histadrut*, published by the Executive of the General Federation of Jewish Labour in Israel, October 1952, pp. 98–103.

233

(3) Membership of the 'kvutza' in the 'Chevrat Haovdim'.

(4) Personal and mutual responsibility of all members of the settlement.

(5) Co-operation in production and consumption, in the education and upbringing of the children; responsibility for the aged and the weak; equality of standards and rights in all spheres.

(6) Property rights must be brought into line with the requirements of the 'Chevrat Haovdim'.

(7) Free selection of members.

(8) Liberty of the individual in all matters of politics and religion.

(9) The settlement of disputes within the 'Chevrat Haovdim'.

The chapter of the Statutes dealing with the 'kvutza' as a unit of the 'Chevrat Haovdim' lays down that the 'kvutza' must be a member of the 'Nir' Corporation, whose function in the agricultural sphere resembles that of the 'Chevrat Ovdim' in other branches of the Federation of Labour's activities. The articles of registration of the 'kvutza' as a co-operative society, which must be approved by the State Registrar of Co-operative Societies, includes a clause defining the 'kvutza' as a subsidiary of 'Nir'. Decisions of an economic nature, such as sale, lease, mortgage or transfer in any other manner of the collective farm can only be effected with the consent of 'Nir'.

Actually, however, 'Nir' exercises this authority rarely and only in crucial junctures in the life of the 'kvutza', as in the case of a split, a liquidation of the Society, etc.

Every 'kvutza' constitutes an autonomous unit:

'Any collective settlement, both independently or in combination for any purpose whatsoever, with any other collective settlements or if affiliated to any national collective federation, constitutes *an autonomous economic unit*: in the selection, acceptance or expulsion of members, in the conduct of its financial affairs, except when two neighbouring and contiguous settlements decide to amalgamate in which case they constitute a single autonomous settlement unit.'

An important clause in the regulations of the 'kvutza' lays down that members are accepted in keeping with the customs of the 'kvutza', as formulated in its supplementary regulations. Any member is at liberty to leave the 'kvutza' after giving prior notice and without causing any damage to the work he performs at the season when he leaves. He may not, however, demand any share in the investment capital of the 'kvutza' or in any other of its collective assets. In so far as 'kvutzot' have special funds accumulated out of profits to improve the conditions of their members, the member who is leaving may demand his share in such a fund.

Other clauses of the regulations are similar to those normally inserted in the articles of association of any co-operative society and deal with such questions as: the rights and duties of members; the rules governing the administration of the society; the status of the general meeting as the supreme body of the 'kvutza'; the rules governing the Federation of Labour's courts; the affiliation of the 'kvutza' to the workers' community; relations between 'kvutzot'; the authority to draft new regulations and to alter or amend existing regulations; the amendment of principles; the partition of the settlement; the liquidation of the collective and the like. In all these the prior consent of 'Nir' must be obtained.

There is little essential difference between the statutes of 'Hakibbutz Hameuchad' and those recently formulated by the 'Ichud Hakvutzot Vehakibbutzim'. Both documents lay down that 'the collective settlement is a large settlement uniting agriculture, industry and handicrafts in its own economy, as well as in outside employment' (Statutes of 'Hakibbutz Hameuchad'). In the Programme of the 'Ichud Hakvutzot Vehakibbutzim' this clause is supplemented as follows: 'in all work at sea or in the air, in public services, and combining work in the public and private economy'. 'Hakibbutz Hameuchad' stresses the 'financial and administrative autonomy of the settlement', while the Programme of the 'Ichud Hakvutzot Vehakibbutzim' is somewhat more detailed: 'Autonomous administration of the settlement and of the social unit upon the basis of the principles of the movement and in keeping with decisions taken by its central institutions.'

The wording of the paragraphs providing for the authority of the national federation over its constituent settlements varies somewhat in the two documents. 'Hakibbutz Hameuchad' declares: 'The planning of the settlement must be discussed and approved by the supreme institutions of the "kibbutz".' The Programme of the 'Ichud Hakvutzot Vehakibbutzim' states on the same subject:

'The authority of the movement over its settlements and members in all matters affecting the functions of the movement and the settlements must be maintained in keeping with democratic principles, and through its competent institutions. The settlements are responsible to the movement and the movement is responsible for the settlements.'

The regulations of 'Hakibbutz Ha'artzi Hashomer Hazair' state:

'Within the bounds of the economic autonomy of each settlement (in regard to its responsibility and administration) the real authority of the "Hakibbutz Ha'artzi" as a general body must cover means of production and consumption, the assets and the profits of the individual "kibbutzim". The authority of "Hakibbutz Ha'artzi" must be expressed in joint responsibility, to secure equal distribution of the burden of the tasks undertaken by "Hakibbutz Ha'artzi", namely: absorption of immigration; a minimum standard of living; equalization of living conditions; education of the younger generation; the consolidation of its settlements; and the general public duties of "Hakibbutz Ha'artzi".'

The authority exercised by 'Hakibbutz Ha'artzi' over its constituent 'kibbutzim' is more comprehensive than that of any other collective settlement federation.

Most of the regulations of 'Hakibbutz Ha'artzi' deal with organizational and economic questions, and do not differ materially from the similar provisions of other collective movements. In the chapter dealing with the objectives and purposes of the 'kibbutz', however, a fundamental difference is apparent.

In the preamble to the statutes of 'Hakibbutz Hameuchad', setting forth its basic principles, we find the following first tenet: 'the co-operative settlement of all members in large and open collectivist settlements'. The regulations of the 'Ichud Hakvutzot Vehakibbutzim' contain the following clause: 'The construction of large collective settlements, actively absorbing immigrants, workers, youth, and children, and exploiting all economic potentialities for the expansion of the economy and the social group, and for the development of the social capacity of the settlements.'

Neither of these two documents deals with political questions. According to the regulations of 'Hakibbutz Ha'artzi', however, the prime objective is:

'(a) common, social and political activity of all "kibbutzim" . . .
'(b) the maintenance of an organic connection between "Hakibbutz Ha'artzi" and the "Hashomer Hazair" World Organization and sympathetic movements for the purpose of joint clarification of all common ideological and educational questions, the absorption of immigration and assistance in providing the immigrants with employment, the participation of emissaries of "Hakibbutz Ha'artzi" in the work of the "Hashomer Hazair" Organization and sympathetic organizations in the Diaspora.'

Paragraph (d) of the second chapter also deals with the question of political activity: 'Determination of the political rights of individuals who demand freedom of political activity rests with the "kibbutz".'

These three clauses dealing with the ties between the 'kibbutz' and a political movement, for common political action, and the manner in which the individual political rights of members are to be treated, provide the constitutional background for the regime of collective homogeneity (or ideology) which is a distinctive feature of 'Hakibbutz Ha'artzi' in all its economic, social and political work.

THE COLLECTIVE WAY OF LIFE

Organization

In the early period of the collective movement all questions even of the most trivial nature were discussed and decided by all members in general meeting, for the membership was small, the problems raised were new and therefore of interest to the entire community. The growth of the 'kvutzot', however, leading to a high degree of specialization and detailed allocation of various duties and functions among the members, made it necessary to appoint committees and even larger bodies to deal with each specific aspect. The general meeting of members is the supreme institution of the 'kvutza', its functions being defined in the regulations, which are approved by the State Registrar of Co-operative Societies. General meetings can be convened whenever it is thought fit, but the annual general meeting must be called after the conclusion of the annual audit of the 'kvutza' accounts, to discuss and approve the annual balance sheet. Two-thirds of the members of the settlement constitute a legal quorum for the annual general meeting the first time it is convened. However, should the first meeting not be held because of lack of a quorum, a second meeting, the decisions of which become valid if passed by a simple majority of members present, may be called. The general meeting elects the Secretariat, comprising all the holders of the principal offices in the settlement, headed by the treasurer, the secretary for outside affairs and internal secretary. A large number of committees are also elected to deal with specific aspects of collective living. Each of these committees is headed by a chairman or a 'Merakez' (executive-secretary). The more important committees are those for the work-roster, the budget, the kitchen, personal affairs of members, culture, school, kindergarten, health, for the various branches of farming and the clothes store. Every 'kvutza' has its yard-foreman and commissary, both of whom fulfil very important and responsible functions.

There are no courts of law in the 'kvutza', no judges, no penalties, no prisons, no titles and, above all, no orders. Visitors from abroad often express their astonishment at the absence of these highly necessary social institutions and the representative of one very important country, who visited a 'kvutza' not long ago, apparently could not believe that the 'kvutza' could operate without a prison, reiterating his question, 'But how can you manage without penalties?'

There is in fact one sanction, imposed only for very grave offences —expulsion. This sanction is utilized only in extremely rare instances, after a highly complicated procedure of inquiry and debate, which must precede the submission of the issue to the general meeting, upon the agenda of which the question has been included. The general meeting alone, by a two-thirds majority of members present, may order the expulsion of any member. Normally, however, matters do not reach this stage, as the person in question will leave the 'kvutza' before the question is put to the vote. A decision to leave the settlement, indeed, is the natural consequence of any serious rupture in the relations between it and any individual member.

Housing

At least four distinct stages can be distinguished in the development of housing accommodation in the collective settlements. The members of the early 'kvutzot' usually adapted whatever Arab buildings they found on their lands—in the majority of cases these were hardly more than hovels in an advanced state of decay. Huts after the Arab model were also frequently constructed. After the First World War, following purchases of surplus equipment from the British Forces, tent camps became the most common form of housing for new immigrants, not only in the 'kvutzot' but also in the suburbs of the larger towns. During the third stage wooden bungalows were constructed, and it is only quite recently that more stable structures, usually of concrete, have been erected. To this day, however, all these types of dwellings can be seen together in many settlements, and the proportion of each, indeed, can serve as a rough indication of the age of the settlement.

The more recent mass immigration into Israel during and after the War of Liberation produced new forms of accommodation, such as huts constructed of timber, light blocks of concrete or canvas. These structures are of a temporary nature and must give way, in the course of time, to other more stable forms of housing. In the main they were intended mainly for the 'Maabarot'—the transitional work camps—but they have penetrated into the collective settlements, too, owing to the lack of capital for building.

A census of housing conducted by the Agricultural Centre of the Agricultural Workers' Organization, on 31st May 1954, covered 212 'kibbutzim', containing 30,338 housing units, accommodating 73,871 persons.

Of the total population in these 212 'kibbutzim' 39,084 were

238

children and 34,787 adults. It must also be borne in mind that 122 settlements were consolidated 'kibbutzim'[1] and 90 were young 'kibbutzim'.

TABLE 89(a)

Adult housing according to types of dwellings

122 consolidated 'kibbutzim'	No. of units	%	No. of occupants	%
Buildings with services	5,414	34·5	10,147	35·5
Buildings without services	5,293	33·5	9,413	32·9
Wooden bungalows, betonade structures	4,952	32·0	9,046	31·6
Total	15,659	100·0	28,606	100·0

TABLE 89(b)

90 young 'kibbutzim'	No. of units	%	No. of occupants	%
Buildings with services	575	12·0	1,198	11·4
Buildings without services	1,303	27·2	2,778	26·5
Wooden bungalows, etc.	2,913	60·8	6,502	62·1
Total	4,791	100·0	10,478	100·0

It will be seen from the above table that in the younger 'kibbutzim' temporary structures predominate—62% of the settlers are still accommodated in wooden bungalows, and only 11% in more permanent structures equipped with the various services.

In the older 'kibbutzim' 35% of the members are housed in properly equipped and stable structures, but this figure in itself suffices to prove the magnitude of the backlog in housing construction. Such accommodation is normally placed at the disposal of older settlers. Indeed only 12% of the settlers who have been members of the 'kibbutzim' for less than ten years live in such buildings.

Five thousand members and parents of members above forty years of age, and 6,250 settlers who have been members of the 'kibbutzim' for ten years and more are still inadequately housed. There are two major reasons for this situation:

(a) The traditional approach among the settlers that farm and other equipment take precedence over housing;

(b) The lack of investment capital and of suitable credit for housing construction, when the settlement emerges from dependence on the Jewish Agency.

[1] The category 'consolidated "kibbutzim"' includes all settlements which have already received the full settlement grant from the Jewish Agency and is used to distinguish them from the young settlements.

Conditions for the accommodation of the children differ fundamentally, though here too there is a certain gap to be bridged.

TABLE 90

Housing of children of pre-school age

Type of building	In 122 consolidated 'kibbutzim'				In 90 young 'kibbutzim'			
	No. of buildings	No. of rooms	No. of children	%	No. of buildings	No. of rooms	No. of children	%
Children's concrete houses	395	1,407	4,254	53	97	352	1,247	45
Children's wooden houses	39	120	391	4	77	251	796	29
Concrete kindergartens	229	1,008	3,969	40	21	89	410	15
Wooden kindergartens	18	60	283	3	24	72	299	11
Total	681	2,595	9,897	100	219	764	2,752	100

The above table shows that in the older 'kibbutzim' 93% of the children are housed in concrete buildings, although it must be pointed out, special buildings for the children have not been erected in all of these settlements and in a number of cases the buildings used for this purpose were originally intended for adults. In the younger settlements, however, conditions are far less satisfactory and 40% of the children are still housed in wooden structures.

A similar situation obtains in regard to children of school-going age as is shown in the table that follows:

TABLE 91

Housing of children of school-going age

Type of building	In 122 consolidated 'kibbutzim'			In 90 young 'kibbutzim'		
	No. of rooms	No. of children	%	No. of rooms	No. of children	%
Concrete buildings	2,482	9,587	79	105	320	65
Wooden buildings	734	2,588	21	58	175	35
Total	3,216	12,175	100	163	495	100

On the average three children occupy a single room, and in cases where the average is four to a room there have been complaints regarding difficulty in preparing homework, reading and study. It will be seen from the table that from the point of view of congestion the situation is better in the younger 'kibbutzim', though a lower percentage of the children than in the older 'kibbutzim' (65% as compared to 79%) are housed in concrete structures.

A normal standard for assessing housing conditions is the area per unit or per capita. In co-operative housing developments in the towns and 'moshavot' in Israel the living area of an apartment intended for four persons averages between 50 and 60 square metres. In the 'kibbutzim' this figure obtains only in buildings equipped with services. It is considerably less in regard to other structures.

Housing units in 'kibbutzim' are constructed as follows: for unmarried persons, one room per person; for married couples—one and a half rooms; for young children—four children to a room; for children of school age—three to four children to a room; for youth and training groups—three to a room. In addition the various communal buildings and services such as the dining-hall, kitchen, clothing stores, cobbler-shop, laundry, culture centre, youth club, clinic, sick room, shower-bath and lavatory, must all be taken into account.

TABLE 92

Housing area per unit and per capita in 212 'Kibbutzim'

Type of building	In 122 consolidated 'kibbutzim'		In 90 young 'kibbutzim'	
	Area per unit (square metres)	Area per capita (square metres)	Area per unit (square metres)	Area per capita (square metres)
Buildings with services for adults	26	13·8	22	10·5
Buildings without services for adults	16	8·9	14	6·5
Bungalows, etc., for adults	12	6·5	12	5·3
Children's houses and kindergartens	24	6·2	24	6·6
Children's houses (for schoolchildren)	24.	6·3	24	7·9
Housing for youth and training groups	12	4·0	12	4·0
Communal services		2·1		2·9

Congestion is far worse in the younger 'kibbutzim' than in the older ones.

The problem of communal buildings is organically part of the general problem of housing in the 'kibbutzim'. In respect of this type of building, indeed, the situation is even worse than in regard to the housing of settlers. In the young 'kibbutzim', all dining-halls, with one exception, are constructed of timber and even in the older 'kibbutzim' 52% of the dining-halls are situated in bungalows. A similar situation obtains in regard to other communal buildings.

The situation in regard to sanitation and hygiene in the younger 'kibbutzim' must give rise to considerable concern. Almost all of the kitchens in these settlements (94%) are housed in wooden buildings, in which, to say the least, conditions are unsatisfactory, adversely affecting the health of those working and taking their meals in them.

Local facilities for hospitalization of the sick are also very meagre.

TABLE 93

Communal buildings in 212 'Kibbutzim' (%)

Type of building	In 122 consolidated 'kibbutzim'	In 90 young 'kibbutzim'
(1) *Dining-halls*		
Concrete	43	1
Wood	57	99
(2) *Kitchens*		
Concrete	53	6
Wood	47	94
(3) *Clothing stores*		
Concrete	36	15
Wood	64	85
(4) *Shoemaking shops*		
Concrete	32	
Wood	68	100
(5) *Laundries*		
Concrete	57	20
Wood	36	65
Planned for construction	7	15
(6) *Culture houses*		
Concrete	39	8
Wood	31	52
Planned for construction	30	40
(7) *Youth clubs*		
Concrete	17	7
Wood	79	38
Planned for construction	4	55
(8) *Clinics*		
Concrete	60	20
Wood	35	72
Planned for construction	5	8
(9) *Sick rooms*		
Concrete	32	10
Wood	7	19
Planned for construction	61	71
(10) *Shower-baths*		
Concrete	59	14
Wood	41	86
(11) *Lavatories*		
Concrete	76	27
Corrugated iron	24	73

In 71% of the younger 'kibbutzim' and 61% of the older 'kibbutzim' there are no sick rooms at all.

The fact that so many of the shower-baths and the lavatories are in wooden and tin structures may cause more cases of influenza and colds and contagious diseases than would otherwise be the case.

It will be seen that the situation in regard to residential and other facilities is still far from satisfactory. The singular structure of social, cultural and economic life in the 'kibbutzim' ought to lead to special attention being given and special efforts made to solve the problems of housing.

NUTRITION

Not so very long ago nutrition standards in the 'kibbutzim' were far below standard. The founders of the movement held that the settlers must make a supreme mental and physical effort and were therefore indifferent to such personal conveniences as decent accommodation, clothing and even food. The comparatively high incidence of sickness which could be traced to malnutrition brought about substantial improvement in this sphere. A contributory factor was, of course, the progress made over the past two decades in the science of nutrition.

Today the 'kibbutz' bill of fare is scientifically planned in accord with modern dietetics. Special cookery courses are organized at frequent intervals with a view to raising culinary standards, and a system of inspection and supervision has been evolved under the control of the Inter-'Kibbutz' Kitchens Committee. Suitable arrangements are made for those who must confine themselves to a special diet for medical reasons.

Many of the settlers keep electric kettles in their rooms to prepare tea or coffee, when meals are not being served in the dining-hall (despite the fact that tea is available at almost all hours of the night or day, for the convenience of settlement guards and others).

Most 'kibbutzim' have installed special cooling devices and cold water is on tap.

Tobacco and cigarettes are allocated to smokers in keeping with their individual requirements.

THE CLOTHING STORES

All work pertaining to the making, mending and care of clothing and footwear is concentrated in the clothing store. At one time the principle of communal consumption was construed as meaning that the settlers were to have no garments of their own. Soiled linen, for example, was handed in and the settler was given clean linen as required. This arrangement has now been changed. Every member may order the clothes to which he or she is entitled under the current 'kibbutz' quota and have them made up to his or her measurements and in keeping with individual tastes. Thus undue uniformity is avoided and the individual is given a considerable range of choice within the limits set by the community.

HEALTH AND SANITATION

Under conditions of collective living, where not only the dining-hall but sanitary installations, too, are communal, the danger of mass infection, especially in the case of epidemic diseases, is always real. However, it is probably not true that the incidence of influenza, for example, is higher in the 'kibbutz' than in the city. An outbreak of influenza is far more apparent in the 'kibbutz', when scores of settlers do not take their places in the dining-hall or report for work. In the city thousands can fall ill without the fact being outwardly noticeable. Nevertheless preventive measures must be far more stringent in the 'kibbutz'. Every 'kibbutz' has its sanitary officer and special sanitary arrangements, but progress in this field is still far from adequate.

All members of the 'kibbutzim' are members of the Workers' Sick Fund of the General Federation of Labour. Every 'kibbutz' has a clinic (and, in many cases, a sick room) in charge of a nurse who is responsible to a doctor who serves a number of settlements in a single neighbourhood. In the larger and more densely populated districts there are central hospitals. The women working in the children's houses also include a number of qualified nurses.

PARENTS OF MEMBERS

At the beginning of the 1953/54 agricultural year there were 2,218 residents of the 'kibbutzim' registered in the category of 'parents and relatives'.[1] The majority of these were aged parents of members of the settlements. The total number of members and candidates for membership in the 'kibbutzim' at the time was 34,508. By dividing this figure by two we can arrive at a rough estimate of the number of family units in the 'kibbutzim', though this figure is very imprecise as it includes grown-up unmarried sons and daughters of members. The category of 'parents and relatives' may comprise approximately

[1] *Statistical Bulletins*. Bulletin A—'Kibbutzim', May 1954. Audit Union for Workers' Agricultural Co-operation Ltd.

12·8% of the total number of units. The following table gives the percentage of parents and relatives in the 'kibbutzim' classified upon the basis of the period in which the latter were founded.

Settlements founded	prior to 1935	14·8%
,,	,, 1936–40	15·3%
,,	,, 1941–47	11·0%
,,	,, 1948–53	9·7%

The higher percentage of parents in the older 'kibbutzim' indicates that the proportion of parents and relatives in the settlement is in direct ratio to its age.

There were 780 males and 1,438 females in the 'parents and relatives' category.

To enable the parents, who for the most part do not share the collectivist ideals of their children, to live their own lives as far as conditions permit, they are housed in a special section of the settlement, with quarters of their own, a 'kosher' kitchen and, if they so desire, a synagogue.

Any desire on their part to engage in some occupation is welcomed, though it is not required of them. They may also work outside the 'kvutza', without any responsibility being accepted by the 'kvutza'. Cases of this kind, however, are quite rare. In general, practically all of the parents who live in the 'kvutzot' together with their sons or their daughters are satisfied with their conditions and many of them are still active despite their advanced age.

CHILDREN

In 1953 the proportion of the total population of Israel under the age of eighteen was 39·1%. Latest data for the 'kibbutzim' show that in October 1953 the percentage of children in this age-group was 38·7% of the fixed population (comprising members and candidates for membership in the 'kibbutzim' as well as parents and relatives). The proportion of children varies with the age of the 'kibbutz' as indicated by the following table.

In settlements founded		prior to 1935	43·1%
,,	,,	,, 1936–40	44·9%
,,	,,	,, 1941–47	37·5%
,,	,,	,, 1948–53	23·2%

The proportion of children in the older 'kibbutzim' accordingly averages approximately 45%.

This situation is reflected in different terms in the computation of the number of children per adult couple (members and candidates for membership) as given in the table that follows:

In settlements founded prior to 1935	1·6 children per adult couple			
„ „ „ 1936–40	1·8 „	„	„	„
„ „ „ 1941–47	1·3 „	„	„	„
„ „ „ 1948–53	0·6 „	„	„	„

In the younger 'kvutzot', it must be borne in mind, the percentage of unmarried persons and of childless young couples is very high. In the older 'kvutzot', on the other hand, families with three or four children are not uncommon.

The economic factor does not influence the size of the 'kvutza' family to any appreciable extent. From the day the mother returns from the maternity home, the infant is placed in a modernly equipped children's house, in the charge of a trained nurse. The mother comes regularly to feed or visit the child. When the infant is weaned the mother visits it in the evenings when she has done her normal work in the settlement. In general parents and children meet in the evenings when the former have completed their work and have washed and changed their clothes. This system is excellent from the educational point of view and constitutes a highly important factor in moulding the character of the younger generation. Ties between parents and their children are very close, the latter being proud of their parents and their function in the 'kvutza'. They seem at least as closely attached to the settlement in which they were born as their parents and in many cases even more so. Even in cases where parents have decided to move elsewhere, the children have often insisted on staying. This attachment was conspicuously demonstrated in recent years, when sons and daughters of the 'kibbutzim' returned after many years on active service during the Second World War and the War of Liberation, or after prolonged periods of study abroad.

In general the children of the 'kvutzot' go to the village school for twelve years. In some 'kvutzot', where the number of children does not justify independent facilities for the upper grades, they attend a district school after their eight years of study. Adequate arrangements are of course made for the pre-school education of the children in kindergartens of various grades.

WOMEN IN THE 'KVUTZA'

The principle of equality as applied in all spheres of life in the 'kvutza' strongly discountenances any attempt on the part of the strong to dominate the weak. The same principle is upheld in the maintenance of equality between the sexes. Nevertheless, consistent implementation has never been easy. The initiative towards enforcing full equality has always been taken by the women and disagreements between the male and female members have not been wanting. In the majority of cases, however, the attitude of the men has been one of indifference. In a large collection of essays, *Chaverot Bakibbutz* ('Women in the "Kibbutz" ', Ein Harod, 1944) one of the contributors writes as follows.

'In the first generation of this collective experiment, not always has the "chaver" evinced a proper understanding of the position of the "chavera", and not always has the latter been able to count upon his support. It is, of course, abundantly clear that this attitude has never stemmed from malice and that factors far more complex were at work—an ingrained conservatism, fortified by the traditions of many generations—but this does not alter the fact nor make it any more palatable. The conquest of labour, the reconstruction of the country, the creation of a workers' society have called for superhuman exertions on the part of all Jewish pioneers, but from the women the effort demanded has been even greater. In the "kibbutz" as a new and unusual, and often very difficult, way of life for the "chaverim", the demands made upon the woman—the wife and mother—have required many special and individual concessions.'

The demands for full equality made by the women were often extravagant in this early period, and they insisted in undertaking tasks far beyond their physical strength in building, road-construction, ploughing, driving tractors and so on. A few women continue to this day to work in these occupations. Time, however, has underlined the over-riding importance of physical ability in undertaking certain tasks, and ability is an important criterion in the allocation of jobs in the 'kibbutz'. There are cases, not very frequent it is true, of men being employed in domestic service branches, either because of a special aptitude or physical weakness. Generally speaking, however, there are more or less clearly demarcated spheres of activity for the men and the women, that of the latter including the kitchen, the clothing store, the laundry, care of children, as well as a number of farm branches, such as the vegetable garden, the nursery, the poultry run and sometimes the dairy.

The major achievement of the 'chaverot' in the 'kibbutz' is, of course, their absolute economic independence.

RECREATION AND CULTURE

The decision of the members of the 'kvutza' to live collectively is sufficient in itself to indicate a high standard of civilization and culture, and indeed the 'kvutzot' constitute a training ground for workers in various cultural fields throughout the State of Israel. They have produced writers, artists, poets, political leaders, military commanders, scholars, teachers and experts in a wide variety of fields. In the collective they insist upon a high standard of culture which the community must perforce satisfy. Where local talent is unable to meet this need, artists and lecturers are invited to visit the 'kvutza'. In all 'kvutzot' there are large libraries, works including belles-lettres, poetry, art, science, history, agriculture, the natural sciences, technology, philosophy, economics, politics, etc., in Hebrew, English, French and many other languages. The collective federations have their own publishing houses and printing establishments, the function of which is to supply the needs of their members for original and translated works.

Cinema shows are a regular feature of 'kvutza' life. Theatrical groups, choirs, orchestras and singers, both of Israeli and foreign origin, frequently accept invitations to appear in the 'kvutzot'. Sometimes such performances are given under the auspices of a number of settlements in a single district. Visits of members to shows and performances in the larger cities also take place. In most areas large amphitheatres, accommodating thousands, have been built jointly by a number of neighbouring settlements.

Any member of a 'kvutza' showing a special aptitude for writing, poetry, art, sculpture or science is assisted to continue his studies in one of the Israel centres, or even abroad. Sometimes special work arrangements are made for such members to enable them to devote several hours a day to their science or art.

In conclusion, it is clear that the cultural needs of collective settlers are met to a far greater degree than members of a similar social stratum in the city.

'NIR SHITUFI'

The 'Nir Shitufi' Corporation has been referred to upon a number of occasions in this chapter, in the course of the discussion of the statutes and regulations of the 'kibbutzim' and the collective federations. The aspiration of the workers' agricultural co-operatives to create a social regime, realizing the principles of justice and equality, provided a powerful stimulus for the development of close ties between the individual co-operatives themselves. In the early years the Agricultural Workers' Organization, a constituent body of the General Federation of Labour, served as the central agency of the collective settlement movement, co-ordinating and directing the activities of its affiliated societies. But the rapid expansion of the work of the Organization itself in a number of spheres, and the maturing of the collective character of the settlements, made it necessary to set up a special body, to ensure maintenance of the principles of co-operation and to foster the growth of a country-wide, planned collective economy. The 'Nir' Central Co-operative Society for Jewish Workers' Settlements, Ltd.—generally referred to in its more abbreviated form as 'Nir Shitufi'—was formed for this purpose.

The objects of the Society as set forth in its articles of association are as follows:

(1) To organize, upon a co-operative basis, workers who are engaged in all branches of agriculture and agricultural training;

(2) To raise the standards of the various branches of agriculture, and to consolidate the farms of its members;

(3) To introduce new workers into agriculture, to assist such workers in their agricultural education and training and to found new farms;

(4) To improve the material and moral conditions of its members, by fostering mutual aid.

All members of agricultural co-operatives are obliged by the statutes of their societies to be members of 'Nir'. Membership is personal. All members are *ipso facto* members of the Agricultural Workers' Organization, and the elected institutions of the two bodies are identical.

The privileges of 'Nir' have been safeguarded in all statutes of workers' agricultural co-operative societies. These special rights include:

(*a*) Compulsory membership of all members of such societies in 'Nir';

(*b*) The acceptance of new members in any society is subject to the prior consent of 'Nir';

(*c*) 'Nir' is entrusted with the task of ensuring the observance of co-operative principles;

(*d*) Should any society go into liquidation, its capital is transferred to 'Nir'.

The statutes of the societies also recognize other rights of 'Nir', such as membership in the society (without membership rights, or the obligation to pay dues or to accept liability for its debts), the right to participate in general meetings of the societies and to exercise one vote, the right to appoint one member of the directorate of the society, etc.

The Audit Union for Workers' Agricultural Co-operative societies is a constituent body of 'Nir'. All Co-operative societies affiliated to 'Nir' are also members of this Audit Union, the function of which is to foster the economic development of its member-societies, to advise and guide them upon the basis of data collected and to audit their accounts.

At the apex of the complex of workers' agricultural co-operatives is the 'Nir' Central Co-operative Corporation for the Settlement of Jewish Workers, in which are united:

(a) Settlement co-operatives: 'kibbutzim', 'kvutzot', 'moshvei ovdim' and co-operative suburbs situated near 'moshavot';

(b) Service co-operatives such as the head office and the branches of the 'Tnuva' Co-operative Societies for Marketing Farm Produce and 'Hamashbir Hamercazi', the central co-operative purchase and supply agency. However, because the latter also serves urban societies, it is not a subsidiary of 'Nir' (which is a purely agricultural body) but is affiliated to the 'Chevrat Ovdim'. In a number of areas the water supply is organized upon a co-operative basis. Such co-operatives are affiliated to 'Nir' and the Audit Union. Co-operative agricultural credit societies are affiliated to the Audit Union of Workers' Credit Co-operatives.

(c) Co-operative contracting corporations: 'Yachin-Hakal'—a contracting corporation formed by a merger between the 'Yachin' and 'Hakal' agricultural contracting societies. The main purpose of this body is to organize agricultural workers on a co-operative basis, thereby to enable them to contract for large-scale agricultural enterprises. The corporation thus facilitates the absorption of new immigrants in agricultural labour in both the private and co-operative sectors of the Israel economy.

Within the general structure of the General Federation of Labour, 'Nir' is a subsidiary of the 'Chevrat Ovdim'. All individual members of 'Nir' and all societies and institutions affiliated to 'Nir' are also members of the 'Chevrat Ovdim', whose competence and rights generally are similar to those which 'Nir' exercises in the sphere of agriculture and cover all fields of co-operation in town and country.

12

AGRICULTURAL POLICY

STUDIES of agricultural economics often refer to the diminishing role of agriculture. From the foregoing chapters, however, it is clear that this does not apply, for the present at least, to agriculture in Israel. Although Israel agriculture has not yet emerged from its first formative stages, it was, from its beginning, directed along the lines of intensification. Agriculture is not intended to occupy a place in the Israel economy similar to that in many other countries. Israel will never be a predominantly agrarian country, with two-thirds of its population engaged in farming. But it does need two-tenths of its population in agriculture instead of one-tenth, as at present.

Before Israel became an independent state, its farmers were guided by an agricultural ideology rather than an agricultural policy. They saw in agriculture the basis for the return of the Jewish people to their homeland.

The basic principles of Israel agriculture, such as national ownership of land; co-operative production and consumption, or both; co-operative buying and selling; cheap credit; organized training, education and research; organized housing; sanitation and medical aid; the application of improved agricultural methods; mechanization, manuring, irrigation, crop rotation, the elimination of pests were all gradually evolved over three generations, and were handed over as a crystallized agricultural policy to the first government of the State at its inception. Ideology had prepared the way for a policy.

The Mandatory Government did not seek to develop an autarchic economy. It utilized its unlimited opportunities as the representative of a great power, and generously supplied, through large imports, all necessary commodities, often without regard to the need to protect local production.

251

It was only during World War II that it became a necessity to restrict imports into Palestine. Although local agricultural production had advanced conspicuously, it was still far from adequate to supply the demand, and imports were never wholly stopped throughout the period. It may safely be said that the state of nutrition was satisfactory in Palestine; better than in some other countries, probably even a good deal better than in England.

Curtailment of imports eventually became necessary, not through lack of currency or purchasing power, but because of transport difficulties and lack of shipping space. In 1942, it became necessary to introduce rationing and controls of certain foods. The High Commissioner enacted the 'Control of Food Ordinance 1942'[1] and appointed a controller with power to issue orders concerning commodities under control. The object of this ordinance was:

'. . . to fix a price, or maximum price, at which each article was to be sold.' (a)

'To prohibit the transfer, or control the transfer of any controlled commodity, from the district where the ordinance is in force—without written permission from the Controller. . . .' (c)

'To prohibit or control the movement of any controlled commodity between districts where the ordinance is in force. . . .' (d)

The system of controls worked fairly well during the war and immediately after. Although the law was infringed now and then by the black market, the situation in this respect was no worse than in other countries where the system of control was in force.

With the establishment of the State, however, a far-reaching change took place. The four months April–July 1948, which constituted the transition phase from the Mandatory rule to that of the State of Israel, was the most difficult period marked by:

(1) A complete land blockade, imposed by the neighbouring countries from the South, East and North, and the severance of these sources of supply.

(2) Impaired sea communications, caused by hostile acts, within Israel territorial waters, and by the decision of foreign shipping companies to cease all sailings to Israel ports or to any ports in the Middle East, where goods could be transferred from one ship to another.

(3) The closing of the Suez Canal to ships carrying Israel-bound cargoes.

(4) Financial difficulties after Israel's exclusion from the Sterling area.

[1] *Official Gazette*, Special Edition, No. 1178. Government of Palestine, Jerusalem, 19 March 1942.

(5) Severance of previous commercial contacts and reluctance of merchants to import in face of great risks.

(6) Lack of personnel for implementing difficult and far-reaching controls.

(7) Diminished local agricultural production as a result of military activity, and the transfer of manpower from production to fighting.

(8) Unluckily, the climatic conditions of the winter 1947/48 were most unfavourable.[1]

All these causes brought about a grave shortage of agricultural produce. Local products rose in price. It became necessary to establish a method of dealing with supplies, production, import, distribution and prices.

The main task of the Food Control Office then set up was to organize immediate supplies.

Steel & Co. Ltd., a foreign firm which was still operating in the country, gave substantial help to the new office. At the same time a consortium of leading merchants was set up, and an agreement was reached for mutual insurance against loss, since insurance companies refused to issue policies on consignments shipped to Israel.

This difficult situation continued for some months, but in August–September 1948 an improvement set in. The sea blockade was gradually lifted. More and more sources of supply were opened up. A system of imports developed, and arrangements for the distribution of fixed and equal rations to all residents was evolved.

IMPORTS

New contacts were now made with international institutions, with governments and with private companies. The supply of grain was assured by the F.A.O. Gradually the supply of foodstuffs returned to normal channels. But it was necessary that the Government should establish itself as the importer of essential commodities, such as grain, flour, sugar, etc. This it did, either directly or through the two bodies, Steel & Co. and the consortium of local firms. Prior to this, imports had been split up among many different merchants. Now it was concentrated, first into the hands of the two companies and eventually in the hands of groups of importers, in addition to the co-operative institutions for supply and marketing of agricultural products.

[1] (a) Report of the Director of Food Division, Ministry of Supply for the period of April 1948–June 1949. Jaffa, 25 July 1949. (b) Report of the Price Controller for the period April–September 1949. Ministry of Supply, Jaffa, 18 October 1949.

The Food Controller took over control of imports by virtue of the order restricting the import trade to 68 commodities.[1] The authorities were careful to maintain a balance between imports and local production. As a result egg imports were stopped entirely at the beginning of 1949 both for civil and military supplies. Potatoes were imported only during the months of local shortage. A striking example of co-ordination was the import of onions, which were not grown locally at first: through price premiums and a flexible import policy, the need was soon fully supplied by home-grown onions.

The control over imports of supplies for livestock, such as grain, bran, oil cakes, proteins, minerals, etc., serves as a useful instrument in directing the development of these branches of agriculture. The Government effectively controls the imports of all equipment for agricultural production. At first, control was vested in the office of Supply, later in the Ministry of Trade and Industry, and in recent years in the Economic Department of the Ministry of Agriculture, which collaborates closely with the Agricultural Department of the Ministry of Trade and Industry. Import plans cover materials and implements for current use and for the extension of agriculture on a country-wide scale. Thus for the year 1953/54 an import programme was planned and carried out to the amount of $43,560,000, of which 45·2% were expended for current production, 20% for fodder and 34·8% for development purposes.

Imported means of production include: pipes and materials for pipe manufacture; equipment for drilling and waterworks; fishing vessels and equipment; agricultural machinery; equipment for dairies, poultry-pens and apiaries; pedigree animals, seeds, fertilizers and materials for their manufacture; materials for fighting pests; equipment, fertilizers and packing materials for citrus exports; equipment for refrigeration, laboratories, etc.

Imports for the year 1953/54 were mostly in the hands of general importers, but the Jewish Agency and the Mekorot Water Company were permitted to deal with their own imports to the extent of $13·2 million, i.e. 34·8% of the total import of agricultural means of production.

CONTROLS

During the concluding period of Mandatory Government, only two commodities were still rationed—flour and sugar. The new control authorities found it necessary to extend the list of rationed commodities. The first order issued by the Food Controller provided for

[1] *Official Gazette*, No. 20, 8 September 1948, Suppl. b.

the control of foods (commodities under control)[1] cancelling the previous order[2] and imposing control over 77 commodities. The list includes almost all kinds of food for human consumption and for livestock and also covers alcohol. Soon it became necessary to add regulations restricting the trade in cattle, sheep and meat.[3] In quick succession there followed orders regulating in detail the trade in various foodstuffs.

With the influx of immigration and the difficulties with foreign currency for imports, the supply position deteriorated and the First 'Knesset' found it necessary to announcet he introduction of an austerity regime at its session of 11th May 1949.[4]

The basic principles of the austerity plan as put forward in the 'Knesset' by the Minister of Supplies and Rationing were as follows:

(1) Reserves of food and other necessities must be assured.

(2) A rational and modest food basket must be fixed, and only those items included in the list being available for the population.

(3) A planned list of articles of clothing, footwear, furniture and household effects, etc., of utility standard would be manufactured and sold at minimum prices, with safeguards to ensure equitable distribution among all sections of the population.

(4) Cost of the rationed foods and utility goods would be reduced in various ways: (a) by imposing charges for less essential commodities, such as white bread, to cover the reduced prices of essential commodities such as standard bread; (b) the sums raised by these additional payments, and the profits on goods imported by the Government itself, would be set aside for this purpose; (c) prices of local agricultural and industrial products would be fixed after careful examination of production costs and services, such as transport, etc., to ensure that the consumer is not made to bear the costs of irrational production, exaggerated profits or inaccurate calculation; (d) the Government would consider the abolition or reduction of customs dues on certain essential articles and also the question of grants by the Treasury with a view to reducing prices of essential commodities.

(5) Production of luxury articles would be allowed only for export and the import of such articles would be discontinued.

(6) In accordance with the requirements of the austerity regime, an import and production plan would be prepared for the period of one year, with priority for utility products, means of production and goods for export.

[1] *Official Gazette*, No. 10, 21 July 1948, Suppl. b.
[2] Ibid., 30 August 1945, p. 773.
[3] Ibid., 28 July 1948, Suppl. b.
[4] *Proceedings of the 'Knesset'*. First Session, Vol. I, Bulletin 10, pp. 486–500.

(7) The Government would promote improved methods for the production of nutritious and cheap foods.

(8) Plans of local production and imports would be regulated in accordance with the following five principles (which may, however, conflict with one another on occasion).

(*a*) Maintenance of an austerity regime.

(*b*) Reduced cost of living.

(*c*) Increased consumption of home-products in order to bring about moderate prices.

(*d*) Saving of foreign currency.

(*e*) Expansion of agriculture and industry for the supply of the local market and for export in order to develop the country and absorb immigration.

(9) The Government would control importers and wholesalers in order to encourage concentration of effort, prevent overlapping and unnecessary costs, and secure an effective wholesale trade.

(10) Charges for services such as hotels, restaurants, cafés, laundries, banks, transport, etc., would be checked with a view to eliminating excessive prices.

(11) Controls in prices, production requirements and marketing would be intensified to give the consumer a fair deal and to prevent expenditure on luxuries. The black market would be fought until it was eliminated.

In determining the national diet, that prevailing in Britain in 1944 was taken as the basis.

The technical implementation of rationing was planned so that each importer should sell his goods to certain wholesalers, determined by the rationing office; each wholesaler was linked to a fixed number of retailers; each retailer in turn linked to a certain number of consumers who bought rationed goods only from him. Ration books were distributed to the entire population.

It is extremely difficult to introduce an austerity regime based on rationing and controls, even where there are strong sanctions to support enforcement, but the difficulties are naturally greater in a democratic society where the sanctions available are limited. An austerity programme of this kind is not popular and it is difficult to secure public co-operation. Defects and failures in carrying out the programme arouse strong criticism and even anger. In Israel, the main difficulty in carrying out rationing successfully was the lack of continuity in supplies. The annual allocation of rationed foods was in fact met, but the monthly rations were often delayed, and at times the housewife had to manage for months at a time without a ration of some particular item. When the promised ration finally arrived, the queuing and time-wasting involved in collecting it further added

256

to her resentment. These justified complaints were supported and exploited to the full by the opposition Press despite the damage to public morale caused thereby. Such circumstances are naturally favourable for the development of a black market. The basis of a black market is of course a real shortage. Plentiful supplies quickly put an end to it, but it flourishes in time of scarcity to which in turn it adds the evils of anarchy. During the period of the austerity regime in Israel, the black market was fed mainly from food supplies brought in by immigrants, gift parcels sent from abroad to relatives in Israel and the sale of rationed items by communities with different food habits (the Yemenites, for instance, were unaccustomed to eating eggs and sold their egg rations). There were also defects in the organization of trade and an excessive number of people engaged in distribution.

Many defects which hampered the control system were due partly to lack of suitable personnel for planning, rationing and control. But the main difficulty in imposing effective controls derived from the conditions of mass immigration and the absence of information beforehand regarding the number of newcomers and the dates upon which they were expected. Shortage and irregular supply of foreign currency created further difficulties.

With the reduced rate of immigration and the extension of local production, both agricultural and industrial, it became possible to release many commodities from rationing and control.

In 1954 controls were curtailed to a large extent, and in agriculture, remained in force on grain, milk and milk products, meat, pond-bred fish and some vegetables (tomatoes, carrots, potatoes and onions) and some kinds of fruit (grapes, bananas).

AGRICULTURAL LINKING SYSTEM

Efficient control of agricultural products should begin with the agricultural producer. This aim can be attained through the appointment of marketing organs, which are given exclusive rights to market the products. The concentration of all products in the hands of such official organs enables the controller to allocate and distribute commodities. The main problem is how best to carry out this concentration of stocks in the hands of these marketing bodies. The problem was first solved in the poultry branch through the agricultural linking system. The transport of fodder from the port to the importers and wholesalers and from them to the producers was put under continuous observation. Merchants applying for licences were required to submit full reports of their stocks.

Produce merchants were permitted to sell to producers only against permits from the Controller. Such permits were given only to producers who marketed their products through official channels and who fulfilled the conditions regarding quantity and quality, as fixed by the Controller.[1]

THE CONTROLLER EXERCISES HIS AUTHORITY THROUGH HIS CONTROL OVER FODDER

This system was first introduced in September 1948 with regard to poultry products. In May 1949 it was given legal sanction in the 'Agricultural Linking Ordinance'[2] issued by the controlling authority, and was extended also to dairy products.

According to this system, each producer has to register at a certain central depot. Each depot must be furnished with a permit from the Controller. The depot serves as a local store for collecting, centralizing and storing agricultural products to be forwarded to the marketing body. The depot and the marketing body are obliged to submit a weekly report to the Controller, of the amounts of goods received and delivered.

The producer is obliged to submit to the Controller a monthly report on the livestock in his possession. Every importer, producer, wholesaler or retailer who imports, exports, deals in, or stores fodder is obliged to submit to the Controller a weekly report of his transactions in fodder and on the stock in his possession.

The Agricultural Linking System has served, and still serves, as an effective instrument for ensuring a fair and rational distribution of fodder among producers in accordance with yield in each branch.

Since January 1950, feed for poultry and cattle has been rationed according to the quantities of milk, eggs and poultry meat delivered for marketing. Fodder is likewise rationed for sheep, draught animals and pond-fish.

The supply of rationed fodder is fixed in accordance with quantities in stock. Locally grown grain is registered and is included in the fodder allocation.

The Agricultural Linking Office functions in co-ordination with the institutions for agricultural planning and imports. In the course of its existence it has proved its efficacy, especially in poultry husbandry. The scope of this branch has varied, during the past six years, the fluctuations being a result of the quantities of fodder allotted each year to this branch. The large quantities of feed

[1] *Proceedings of the 'Knesset'.* First Session, Vol. I, Bulletin 10, p. 401.
[2] Agricultural Linking Ordinance published in two dailies, *Davar* and *Al Hamishmar*, 25 May 1949.

required by poultry-farming come mainly from imports, and these are under control.

The following table reflects the fluctuations in imports of fodder, which have in turn brought about variations in the scope of poultry-farming:

TABLE 94

Imports of fodder (tons)

	1949/50	1950/51	1951/52	1952/53	1953/54
Grains	99,885	95,742	43,454	65,937	31,216
Carobs	21,391	19,374	9,043	10,876	3,297
Bran	12,358	8,891	3,152	2,027	2,483
Oil cakes	26,059	30,079	12,683	13,690	23,249
Meal of meat					
and fish	13,791	10,536	6,116	7,475	10,013
Milk powder	—	—	125	200	—
Vitamin powder	—	—	—	201	38
Cod liver	370	244	20	69	600
Bone meal	—	—	1,867	700	1,743
Lupins	3,530	1,667	1,598	9,715	6,600

Imports of the principal kinds of fodder: grain, carobs, bran, oil cakes and animal protein, reached the highest level in 1949/50. That year also marked a maximum yield of poultry products. Imports of grain were lowest in 1953/54, as in that year local yields were exceptionally high. But the reduced imports of oil cakes and proteins prove that this year, too, there was a recession in the poultry yield, as compared with 1949/50. This same tendency can be observed in each of the four years following the peak year of 1949/50.

At the beginning of 1954, the Ministry of Agriculture decided to abolish planning for the poultry branch, as such a policy requires a reliable supply of fodder. Since then the scope of poultry-farming has been regulated by the allocation of foreign currency.

A new schedule has been fixed for the allocation of fodder, according to which 92 grams of concentrates is given for each egg and 1,050 grams for each kilogram of poultry meat marketed.

In September 1954 it was decided that instead of raising prices, an experiment be made in allowing the free marketing of all surplus eggs above the 25 million eggs a month required for rationing. The price of eggs in the free market did not reach excessive levels and it was decided to continue with the experiment. Prospects in this branch seem good. Having first begun with the Agricultural Linking System, the authorities are now discussing the future of poultry-farming with a view to abolition of controls and of subsidies for feed, and the development of poultry-farming as a main source of supply for meat, and probably also of eggs for export.

PRICES

Methods of regulating prices for agricultural products have varied from time to time. In 1943 the Mandatory Government set up a board of experts who advised taking production costs as the basis for official prices. This principle was implemented, more or less, during the period of World War II. But in the first months after the founding of the State, it proved impossible to find a satisfactory basis for price calculations. It was necessary to deal first with the means of production, in order to be able finally to fix prices of the finished product. It so happened that in the winter of 1948/49 prices of fodder were falling continuously in the world market, and each shipment was cheaper than the previous one. Merchants found it inadvisable to lay in stocks of which there were already considerable quantities in the country. At this time there was a strong tendency to expand poultry-farming. At first the authorities saw no necessity to deal with prices of fodder, as the supply exceeded the demand, but as stocks diminished it became necessary to fix prices without delay. Livestock-farming—poultry and dairy—occupy a very important place in Israeli husbandry. It was, therefore, very important to direct prices for the products of these two branches. The Price Control Office adopted the method of advance price fixing when issuing import licences, with the right to re-examine prices when the goods arrived in port. The same method was adopted for the import of fertilizers. The profits allowed to the importer varied from 6% to 11%; and for fertilizers a profit of 6% was fixed.

Gradually this method embraced also seeds, agricultural machinery, spare parts, pipes and other items.

In recent years, methods of fixing prices for means of production have been evolved. Sales prices are uniform for all articles of the same kind, but purchase prices vary in accordance with the country of origin and fluctuations in world markets. In order to maintain a uniform price, importers are required to pay sums equivalent to the difference between the price fixed for each consignment and the purchase price—after deducting their expenses and profits. In this way Equalization Funds were established for various means of production.

As prices of fodder fell, considerable amounts accumulated in the Equalization Funds. In the years 1954–55 the Ministry of Agriculture utilized these funds mainly for experiments in the export of agricultural production.

Prices for both imported and locally made pipes are officially fixed, although rationing of pipes has been abolished as a result of market stability in this commodity.

With regard to agricultural machinery the method is as follows: For a tractor costing less than $5,000, an additional sum of 19% is added to the price. This includes the importer's profit, services and one year's guarantee on the machine. If the cost is higher than $5,000, the total addition is 15%. On all other agricultural implements the addition is 10%. Besides the above payments another 3% charged on all implements is paid into a special Equalization Fund for agricultural machinery.

Other materials under control are iron, timber, tin-plate, corrugated iron sheets, motors, generators, thread, sacks, etc.

Mixed farming is fairly complex and is full of problems, contrasts and contradictions. Arab farming is extensive while Jewish farming is intensive. The old-established farming settlements suffer from a shortage of manpower, and the newly established farmers have not had time to acquire efficiency and expert knowledge. In dry farming differences are considerable in various areas owing to climatic conditions. The size of lots under irrigation varies considerably and many farms have not yet got a minimal area under irrigation. The structure and equipment of farms vary likewise. Vegetables and fruits are greatly affected by seasonal changes and, finally, the fate of a season's crop may be determined by a heat wave. In these circumstances it is difficult to define the limits between development costs and costs of seasonal production, and therefore not easy to fix prices.

During the period of extreme shortage the system of price controls was aimed at protecting the consumer. But in recent years, with the expansion of agricultural production, control is intended for the protection of the producer.

At first the attitude towards vegetables and fruit prices was opportunist, rising and falling with the seasons. In June–August 1950 an attempt was made to decontrol prices of tomatoes and cucumbers and to announce their market prices over the radio.

In 1951 an attempt was again made to fix subsidies according to a schedule agreed upon between farmers and the administration in order to increase production. But this plan failed, as a result of the inflation prevailing in that period when prices fixed in December 1950 were no longer realistic in June–July 1951.

More flexible methods of price fixing became necessary. Maximum and minimum prices were fixed. New settlers were guaranteed minimum prices for essential vegetables and to this end a Minimum Price Fund was established. The fund was financed by deductions from the proceeds of each ton of vegetables reaching the market plus a Government contribution of 25% to 35%.

The authorities could not find an effective system of control over cereals and price fixing for grain was therefore abandoned.

In summer 1952 it became possible to de-control prices of agricultural products, and at the time of writing, in spring 1955, three agricultural branches alone remain under control: Poultry, dairy and fish-ponds.

With the expansion of agricultural production and with the comparative stabilization of the Israeli economy in general, no marked changes have occurred in prices of agricultural products since July 1953. It should be added that the Government still calculates prices for fodder and fertilizers at the low exchange rate of one Israel pound per dollar. With regard to some products, such as cauliflower, potatoes and bananas, official prices were fixed in 1951, by means of a subsidy, mainly in order to lower the prices for the consumer and to encourage the use of summer potatoes.

In the event of a glut of vegetables, the Surplus Board, affiliated to the Minimum Price Fund, takes over the surplus against payment of minimum prices.

The price of milk has not been changed since July 1953 and is 254 pruta per litre to the producer and 38 pruta to the milkman. But the consumer pays only 250 pruta per litre. Since then costs of production have risen, in spite of greater efficiency and higher yields. The slaughtering of cows resulting from this state of affairs is a cause of great concern. A new calculation has therefore been made on the basis of production costs of 286 pruta per litre instead of 254 pruta. The following are the items of expenditure per annum for a cow yielding 3,800 litres: Feed: £I.571; labour: £I.207·760; amortization: £I.87; maintenance of bull, insurance and veterinary aid: £I.64; straw: £I.15; sundries: £I.15; general expenses and insurance: £I.164·780. Total expenditure £I.1,124·540. Income from calf or bullock and manure £I.116. Final total expenditure £I.1,008·540, or 265 pruta per litre. To this should be added 20 pruta for refrigeration and transport, and 1 pruta for an anti-tuberculosis fund. Thus we arrive at production costs of 286 pruta per litre.[1] After September 1951, the Government retained the low price of milk for the consumer through subsidies and by mixing fresh milk with milk powder, in varying quantities: from 25% at the beginning of the period, down to 4% at the end of 1954. The standard milk thus obtained reached the population without deterioration of quality.

[1] Report of the Economic Division, Ministry of Agriculture, for the period 1 October 1953 to 30 September 1954, p. 35.

Since summer 1953 the production of cream has been permitted. This was welcomed as cream is very popular. The price of cream was fixed by the dairies, with the approval of the appropriate Government offices.

Prices of poultry products to the consumer were, in 1954, 70 pruta per egg and £I.2·400 a kg. of meat, on the average. For the year 1955, a new average price of 76 pruta per egg for the consumer has been set, of which 65·7 pruta goes to the producer—as against 57 pruta in 1954.

Subsidies are being granted for products intended for export. In deciding to lift control from any commodity, one consideration is whether that particular commodity is included in the basic food basket which determines the cost of living index. If that commodity does not affect the index, de-control is more easily accepted. Although present production conditions allow for the lifting of price control on many commodities, the Government has not abandoned the principle of price control and reserves the right to revert to this policy should the need arise.

SUBSIDIES

The purpose of subsidies is to secure the farmer a full return for his day's work in case the price of his product does not cover his expenses. (The gap may be due to unfavourable conditions or to insufficient experience in agriculture.)

Actually, the use of direct subsidies to producers has been very limited. In the course of three and a half years the total amount of such subsidies did not reach the sum of £I.4 million. The highest amount was paid out in the peak year of 1953/54 for vegetables and olives.

The main purpose of subsidies is to encourage the cultivation of industrial plants such as ground-nuts for oil and for export; cotton, sugar-beets, tobacco; also tomatoes for pulp, cucumbers for pickling, sweet-potatoes for starch, etc. The subsidy for tomato-pulp for export was high. In 1954 production costs were some £I.100–£I.110 per ton, whereas the factories paid only £I.60 per ton.

The Government subsidy for milk ranged from 14 pruta per litre in December 1951 and November 1952 to a maximum of 49 pruta in April 1953, and at the end of 1954 stood at 38 pruta per litre. There were instances of subsidies paid for ploughing in the Negev. Subsidies were also paid for exported products, in cases where the producers might not have exported had they been allowed to sell their products on the home market.

The policy in Israel is to limit subsidies to the producer as far as possible. Where a form of subsidy is required on grounds of general economic policy, it is thought best to maintain prices at a fixed level by grants to reduce the prices of essential commodities. The budget for 1955/56 includes a total amount of £I.42,250,000 as grants to reduce prices of essential commodities. The details are given in the following table:[1]

TABLE 95

Government grants in 1955/56

	£I.	£I.
(1) *Exchange rate differences*		
(a) Food	17,500,000	
(b) Health	2,400,000	
(c) Agricultural production	11,100,000	
(d) Culture	500,000	
		31,500,000
(2) *Subsidies to agricultural produce*		
(a) New settlements	1,500,000	
(b) Milk	4,500,000	
(c) Ground-nuts for oil	1,700,000	
		7,700,000
(3) *Reserve*		3,050,000
Total		42,250,000

Of this considerable sum, £I.7,700,000 or 18% is allocated to agricultural production. This also includes the subsidies which are given to the consumer and not directly to the producer.

The Government is of the opinion that should this grant be withdrawn, the C.O.L. index would rise by some 13 points.

Subsidies for agricultural production include the following:

(1) For new settlements—according to contracts between the Ministry of Agriculture and the marketing bodies.

(2) For milk—through the marketing bodies and the milk depots.

(3) For ground-nuts—through the Controller, in accordance with receipts from the oil presses for quantities of nuts delivered.

(4) For fodder, when credits have been opened.

FUND FOR MINIMUM PRICE INSURANCE

This fund was set up by the Economic Division of the Ministry of Agriculture, at the end of April 1952. Its Directorate consists of

[1] *Introduction to the Budget of income and expenditure for the financial year 1955/56*, Jerusalem, January 1955, p. 279.

twelve members who represent the marketing bodies, the producers and the Government. The Directorate of the Fund meets fortnightly to take decisions in accordance with prevailing market conditions. The Minister of Agriculture may veto decisions of the Directorate within ten days after their adoption. The Directorate of the Fund is authorized to fix the amounts to be charged for the Fund on various kinds of vegetables, in accordance with prevailing market prices; also to fix minimum prices in periods of surplus; and to deal with surpluses by means of a special board.[1]

The purpose of all these policies is to promote the expansion of agricultural production, to increase the numbers of agricultural producers and to supply the population with a suitable food basket. These methods may not always have been as effective as was hoped, but on the positive side it must be borne in mind that in the course of six years agricultural supply was doubled, imports decreased and exports increased, and prices of agricultural products have remained stable for the past two years in spite of inflationary pressures. All this testifies that the Government's economic policy and organization has had a marked effect on Israeli Agriculture.

Measures of this kind are usually known as a progressive agricultural policy. Their importance is all the greater against the background of Israel's agrarian policy.

AGRICULTURAL CREDIT

No systematic study of the problems of agricultural credit in Israel has so far been made. For this reason our inquiry into the question perforce bears an empirical character, and is based largely upon the personal experience and impressions of persons directly involved. Indeed, it has proved impossible for us to secure exact and authoritative data regarding extensive categories of borrowers—and even lenders—while even such material as could be collected, at the cost of much effort, suffers from the prevailing lack of system and uniformity. For example: the financial year is not fixed. For some institutions it closes at the end of December, for others at the end of March. Even the term 'agricultural credit' is not unambiguously defined. Not infrequently settlement and marketing agencies are included in this category. And finally, as if the task of analysis and collation was not sufficiently complex, the student has to reckon with the fluctuating value of currency, particularly in the past five or six years.

The data available in this field are woefully inadequate, and whatever conclusions we may venture to draw are qualified by this lack

[1] See Note 1, p. 262 (Report, etc., p. 30).

of information. It is a veritable labour of Sisyphus to attempt any comparative analysis of credit conditions in Israel and those obtaining in other countries. Studies made abroad refer to agricultural economies with a long tradition, where initial investments are lost in the mists of generations long past. In Israel the first generation of pioneers has not yet consummated its task. Like a tree in the early phase of its growth it must strike root deeper, and put forth new shoots and branches. There is no real parallel, in the United States, or Soviet Russia or any other country, to these settlers—neither the veterans who in the course of twenty-five or thirty years have paid off their debts to the colonizing institutions, nor the newcomers, who are still bound by their contracts. Indeed it is because finite chronological bounds can be put to it that this problem is so absorbing and instructive a subject of study and research.

ORGANIZATION OF AGRICULTURAL CREDIT

Even without delving any deeper it is clear that basically agricultural credit in Israel suffers from defective organization.

In more advanced countries a special administrative body controls agricultural credit. Thus in the United States the Farm Credit Administration is an independent agency under the authority of the Secretary for Agriculture. The Administration has twelve district offices, comprising four main credit units operating under a single roof. They are: (1) a Federal Land Bank, extending long-term loans secured on mortgages, in the majority of cases through local farm credit societies; (2) a Production Credit Corporation, which establishes and administers producers' credit societies in the district in which it operates; (3) a district Bank for Co-operatives, granting credits to farmers' co-operative societies; (4) a Federal Intermediate Credit Bank discounting 'agricultural paper' for co-operative societies, banks, etc.

These banks do not deal directly with private applicants for loans. They operate through various agencies—marketing and producers' credit institutions and the like. The basis of this credit is 'agricultural paper', in other words bills duly accepted by the farmer and endorsed by one or other of the institutions referred to.

This of course does not complete the list of institutions granting agricultural loans. Insurance and farm mortgage companies, which conduct their business on an ordinary commercial basis, are also active in this field. Nor have credits of a usurious nature, which for time immemorial have battened upon the tillers of the soil, been completely eliminated. The clientele for this type of loan comes mainly from farmers forced to sell as a result of their own incompetence, who because of the state of their farms are unable to enlist bankers' credits. In these cases, the lender often plays the part of broker and arranges the necessary credit for the seller, secured by a mortgage or otherwise.

Agricultural credit is protected by public law, and is controlled by the State to ensure its constructive character.

In some countries a distinction is drawn between agricultural credit proper and 'social' credit. The former is administered on a business basis, taking into account such considerations as marketing, price levels, interest rates, the dates upon which bills fall due and the like. 'Social' credit, on the other hand, is more in the nature of farm relief and the purely economic aspects of such loans are not of primary importance. They are given, generally, when the State wishes to make good the difference between high costs of production and low prices, or between disparate interest rates.

Diverse forms of farm credit are taking shape, as the State regards this as one of its most effective agropolitical instruments in promoting agricultural development.

The early Zionist colonizers did not ignore the question of agricultural credit. Theodor Herzl, founder of the Political Zionist Movement, floated the Jewish Colonial Trust, to facilitate his efforts to secure a charter over Palestine from the Turkish Sultan. The necessary capital, however, was not subscribed and the charter was not granted. Nevertheless, the Jewish Colonial Trust was successfully floated and subsequently founded the Anglo-Palestine Bank (today the Bank Leumi L'Israel), the largest financial institution in Israel. However, the Anglo-Palestine Bank did not enter the field of agricultural credit at all for the first thirty-four years of its operations. The task of financing agriculture devolved upon the colonizers themselves. Baron Edmond de Rothschild, who engaged in colonization on a comparatively large scale, was the first in this field, and in addition to founding settlements supported the settlers materially. The activity of the Zionist Organization in this sphere was strictly circumscribed by its meagre funds, and it founded new villages at the rate of one, two and even three in a year, though there were lengthy periods of years in which not a single settlement was established. The settlements it sponsored were given their equipment in dribs and drabs over a period of many years—at times ten and even more—instead of the optimum of two or three. Agriculture was not sufficiently mature to have recourse to commercial banks. The colonizing institutions were compelled to create credit institutions, but the capital at their disposal was in no way commensurate to their needs, and from that day to this inadequate credit has been the bane of Israel's farmers.

Three types of agricultural credit

In Israel agricultural credit is normally classified in three categories: (1) Settlement credits, granted by the colonizing agencies, on a long-term basis—25-30-40 years—upon easy terms and at low rates of interest; (2) long- and medium-term loans provided by the Jewish Agency, the Government and other public institutions; (3) short-term credits, usually granted for a term of months and under no circumstances for more than one year. This type of credit is given mainly by banks and suppliers.

Upon the basis of available data, the following table reflecting the volume of each of the three types of agricultural credit has been compiled.

TABLE 96

Composition of agricultural credits in Israel (£I. millions)

Year	Jewish Agency allocations	Other long- and medium-term credits	Short-term credits Banks	Others	Total credit
1948	10·0	4·5	2·0	0·8	17·3
1949	17·2	8·4	4·8	2·5	32·9
1950	33·8	13·2	9·2	5·5	61·7
1951	51·0	19·5	12·9	8·8	92·2
1952	78·8	29·2	18·0	15·1	141·1
1953	116·7	39·7	21·0	12·5	189·9

The large sums allocated by the Jewish Agency in the past two years have been invested in the consolidation of existing settlements. Comparatively small sums of money were invested in the foundation of new settlements. The main—probably the only—source of long-term loans at present is the Government Development Budget, which is largely administered by the Israel Agricultural Bank, a State institution founded in 1951.

Table 96 shows the steady expansion of the volume of long- and medium-term credit, during the five year period under review. In Table 97, which follows, the increasing volume of farm credits is

TABLE 97

Growth of credit, investment and value of production
(£I. millions, current price levels)

Year	Growth of long- and medium-term credits	Agricultural investment	Value of agricultural production
1949	11·1	14·0	34·6
1950	21·4	26·0	43·8
1951	23·5	30·2	55·7
1952	37·5	40·3	121·8
1953	48·4	54·4	190·0 (estimate)

compared with annual agricultural investment and production totals.

Investment in agriculture in this period aggregated £I.165 million, while credits extended to finance investment totalled £I.142 million. The difference of £I.23 million represents settlements' own capital, as well as credits obtained from unspecified sources.

Integration of the diverse forms of credit

Neither the bounds of each form of credit, nor the purpose to which it is put, can be clearly determined beforehand. Investment capital is not infrequently used as working capital to finance current production, but more often the situation is reversed and working capital, secured for very short terms, is utilized to eke out insufficient investment capital.

As long as agricultural production in this country lags behind consumption, farmers in both old-established and new settlements will be eager to expand existing branches of farming and to develop new ones as the need arises. Thus a great and growing demand for credit to finance agricultural production on the lines envisaged in development plans may be expected in the next few years.

Assuming that the economic problems confronting agriculture in Israel are solved and projected development proceeds in keeping with available investment capital, we may expect expansion to be on the following lines: New farm-units will be developed within the limits set by the funds at the disposal of the colonization authorities. The settlements themselves will continue to expand even after the consolidation stage has been reached. Farm branches will be developed as funds for further investment are secured. The credit problem will then recede to the volume required for working capital and short-term credits.

The optimum ratio between the volume of working capital and the value of current annual production has been variously estimated. It has been placed at 2 : 5, 1 : 3 and 1 : 4, based upon the experience of other countries. But here, too, there is no real common standard for comparison, as farming in the temperate zone, with its long hibernal season, differs diametrically from conditions in Israel, where operations continue uninterrupted throughout the year. In Israel as elsewhere, each crop has its due season, but hardly a week passes, in winter as in summer, without a harvest of some sort—of grain, of fodder, of vegetables or of fruit. Every crop passes through three principal phases: (*a*) of preparation, ploughing, seeding, planting, etc.; (*b*) of vegetation, when the crop is entirely at the mercy of the elements; and (*c*) the season of harvest, vintage, picking and packing.

Credit requirements vary from season to season, in keeping with the composition of the crops and the time that must elapse before they are converted into cash.

In economic discussion, it is always wise to avoid averaging, for the margin of error is wide. In the present instance, however, there is no alternative, while the range of deviation is not extensive as we are dealing in terms of thousands of farm-units and tens of millions of pounds.

An analysis of 212 'kibbutz' balance sheets

We have before us a digest of 212 'kibbutz' balance sheets. The settlements are classified according to age in four categories: (1) those founded prior to 1936; (2) founded 1936–40; (3) founded 1940–48; and (4) founded 1948–52. We shall examine the relation between fixed assets, value of securities and current assets and subsequently the relation between loans for more than twelve months, working capital and sundry creditors. Finally we shall study the relation between total borrowed and own capital.

	Category 1 40 settlements		Category 2 53 settlements		Category 3 43 settlements		Category 4 76 settlements	
	1950	1952	1950	1952	1950	1952	1950	1952
Fixed Assets	66·6	62·8	73·0	71·1	74·0	73·3	72·8	73·0
Securities	6·9	7·0	4·8	5·5	3·2	3·4	2·7	3·3
Current Assets	26·5	30·2	22·2	23·4	22·8	23·3	24·5	23·7
Total	100·0	100·0	100·0	100·0	100·0	100·0	100·0	100·0
Loans for more than one year	49·4	44·9	65·7	60·6	70·1	67·5	75·5	71·6
Loans for working capital	26·6	23·5	18·4	19·3	15·2	16·4	9·9	12·3
Sundry creditors	13·9	18·3	11·5	14·7	10·9	12·4	10·8	12·2
Total borrowed capital	89·9	86·7	95·6	94·6	96·2	96·3	96·2	96·1
Total own capital	10·1	13·3	4·4	5·4	3·8	3·7	3·8	3·9
Total	100·0	100·0	100·0	100·0	100·0	100·0	100·0	100·0

Initial capital exceeds 70% in the younger 'kibbutzim', but falls short of this porportion in the older established settlements. In regard to current assets and securities the situation is reversed. These securities are not purchased at the initiative of the settlements. They represent shares in the various Funds of the Movements to which the settlements are affiliated, in marketing agencies and in financial institutions which grant loans only to shareholders. The younger 'kibbutzim' invest less of their capital in these securities, their aggregate being approximately half of that of the older settlements.

The composition of outstanding indebtedness reflects the age of the settlement. The volume of investment and of medium- and long-term credits are closely related. This is particularly clear in the figures for 1952, when additional allocations were made out of the Jewish Agency and the Israel Government Development Budgets. It is also reasonable to suppose that the older-established settlements were given more liberal credit by their suppliers, which explains the higher percentage of sundry creditors in their balance sheets.

The shortage of working capital which has become chronic throughout Israel in the past two years underscores the difficulties experienced by the younger settlements. In the older settlements, it will be seen, working capital constituted 19%–26% of the total, in the 'consolidation' settlements it was 15%–16%, in the younger settlements it did not exceed 13%.

We shall not venture to fix any optimum relation between the volume of working capital and the annual value of production, as this is a subject requiring much closer study. Nevertheless, we assume that this ratio should be more or less the same for all settlements. This assumption is supported by persons familiar with the problems and conditions involved.

Credit in relation to production

We must rely upon the figures cited earlier in this chapter as reflecting the financial state of affairs in agriculture. The following table shows the relation between working capital and the value of production in the years 1949–53.

Year	Percentage
1949	14
1950	21
1951	23
1952	15
1953	11

In the course of the past two years the situation has deteriorated. It is reasonable to suppose that in 1953 working capital of a magnitude of £I.38 million, namely 20% of the total value of production (£I.190 million), would have eased the position of the settlements to a very substantial extent; if this proportion were increased to 25% the problem would probably have been liquidated entirely.

Main lines of credit policy

The question is not restricted to the purely technical aspect of mobilizing more funds for this purpose. Anti-inflationary policy implies limitation of the volume of credit—agricultural credit as well as any other. An increase of £I.20 million in the aggregate of agricultural credit would certainly aggravate inflationary tendencies. We are, however, of the opinion that the situation can be mitigated, at least, by measures of an organizational character, which would reduce the amount of the new money required. An increase of £I.20 million in farm credit would reduce the burden resting upon the financial secretaries of the settlements, who are at their wits' ends trying to make ends meet. But this relief would be very short-lived, and retribution, in the form of an inflationary rise in price levels, would not be far behind. It would soon transpire that the increase in credit was inadequate. As a short-term, emergency measure, an increase of ten million in the volume of farm credit might be permissible provided that simultaneously further-reaching measures were set in motion. It is our opinion that agricultural credit policy, even in the near future, must be integrated within a more comprehensive policy.

The main lines of such a policy, we believe, must be as follows:

(1) Real investment must go hand in hand with the extension of credits by the colonization and development institutions. The present frantic efforts being made by the settlements to secure interim loans to tide them over 'until our allocation is approved' must cease. In the past, it is true, we worked—and taught others to work—according to this system. At one time, and under other circumstances, it might have expressed a certain audacity, even vision. Today, however, when we must reckon in terms of tens, and even hundreds of millions of pounds, methods more attuned to objective conditions are essential, in the ultimate interests of the agricultural sector itself. We must not forget the frustrations and failures that have bestrewn our path in settlement on the land, the individual and even mass desertions of newly established villages, the waste of manpower and equipment—all because of precipitate haste and the absence of minimum preliminary conditions. We must have the courage to admit our own culpability which inevitably led to that cardinal sin of our work of land settlement—desertion of farming, and an efflux to dead-end jobs in the city, and even, upon occasion, to emigration from Israel.

(2) Settlements must accustom themselves to using loans within their limits and for the specific purpose for which they are granted. Thus a loan of £I.1,000 may be utilized to finance an investment of a magnitude of £I.1,200, but not one of £I.2,500; otherwise the borrower will find himself floundering outside his financial depth. If he is convinced of the necessity of a £I.2,500 investment, then he must secure at least £I.2,000. Some instrument of control and supervision must be established to ensure compliance with these principles.

(3) Investment loans, with the exception of credit extended by the Jewish Agency, must be administered by a single authority. Under present circumstances the situation is approximately as follows: Bank A gives a loan repayable in three years to the extent of one-half —let us say—of the value of a tractor or combined harvester, or whatever else it is that the borrower wishes to buy. Then Bank B grants a one-year loan, amounting to one-fifth of the purchase. Both loans together do not cover the cost of the equipment, so the borrower must mobilize capital of his own to pay the balance outstanding within the required term of three months. And then three months later another loan is obtained from 'a private source' at 18%–24% (and even more), so as to avert default on the debt. Why cannot Bank A grant up to two-thirds, or three-quarters of the value of the investment, or if its funds are insufficient why cannot it reach some arrangement with Bank B to provide the sum required? Now that Israel possesses so efficacious an instrument of agricultural policy as the Agricultural Bank, which in the five years of its operation has had a turnover of a hundred million pounds, surely that little host of 'public funds' with their puny capital, maintaining themselves by renewing loans and collecting interest, are redundant. It is only too obvious that the activities of these institutions scotch all attempts at the rationalization of agricultural credit.

(4) We must adopt the system of prior ordering of certain crops through the agency of the Ministry of Agriculture, upon the basis of contracts with the primary producers, along the lines incorporated in the British Agricultural Act of 1947, so successfully administered by the Government during the period 1947–52. Under this arrangement the institution ordering the crop makes an advance payment to the producer through the marketing organization. The choice of crops may be extended or restricted in keeping with needs and prospects. The first list of crops under such a scheme could include industrial crops, such as fibre crops, oil, sugar, as well as calves for slaughter and perhaps also legumes and fodder. Extension of the scheme to include a wider selection of farm produce in suitable quantities would provide a substantial measure of relief for the producers without adding to their financial burdens. Introduction of this system would also help to reinforce marketing through organized channels.

(5) An Agricultural Credit Authority attached to the Israel Bank of Agriculture, and comprising representatives of the Jewish Agency, the Ministry of Agriculture and the commercial banks must be created. The Authority must co-ordinate the work of the banks operating in the agricultural sector. It must also act in liaison with the Ministry of Finance and the Knesset Finance Committee in the

implementation of financial policy, in executing Government guarantees and in laying down the lines of agricultural credit policy.

(6) Another measure which could concentrate short-term bankers' credit would be to strengthen relations between any settlement and a particular bank. In more advanced countries farmers have their own regular bankers, with whom they conduct all their financial transactions. Here in Israel, in the balance sheet of one 'kibbutz' no less than *nineteen* financial institutions with whom it does business are listed, not counting the two or three co-operative societies and the private businessmen which would swell the list of creditors by at least an equal number. This adequately illustrates the necessity for concentrating investment credit in one institution and loans for working capital in another—if for any reason it should be found that both branches of business cannot be conducted under a single roof. But a precondition of success is systematization and standardization of banking methods and prevention of a situation in which there are—from the borrower's point of view—good bankers and bad. The projected Agricultural Credit Authority must ensure equality of treatment for all clients, and must keep the volume of credit related to the value of current production, or to whatever other index might be found suitable.

(7) More suitable arrangements for the purchase, storage and financing of supplies must be evolved. Probably the creation of a National Supply Council comprising representatives of various national and regional supply institutions, of the proposed Agricultural Credit Authority and the Ministry of Agriculture, would best serve this purpose. By establishing a network of district supply depots, the council could do much to alleviate credit problems, by reducing the burden of current debt under which so many settlements labour.

(8) As long as the German Restitution Payments Agreement remains in force, all imports of agricultural equipment should be made through the agency of the Agricultural Investments Company Ltd., working in close collaboration with the Ministry of Agriculture and the Ministry of Finance. Such an arrangement would reduce the volume of currency in circulation, without aggravating the dangers of inflation.

(9) The 'consolidation' settlements must be given preference in the allocation of funds for development, without neglecting opportunities for an expansion of production in the older villages.

(10) In respect of working capital, too, the 'consolidation' and new settlements must enjoy priority—in keeping with the volume of current production—in order to bring their conditions into line with those of the older-established villages.

274

(11) Representatives of all bodies and institutions interested in farm credit, including the financial secretaries of settlements, the banks, the Ministry of Agriculture, the Ministry of Finance and the agricultural organizations should meet annually to discuss common problems and to exchange information.

(12) Research must be initiated into various economic problems of agriculture, including farm credit, by some independent Institute of Agricultural Economic Research, attached to the Ministry of Agriculture, the Hebrew University or the Agricultural Experiment Station. In the event of agricultural research and instruction being unified under the aegis of the University then such research should be located within a University Institute for Agricultural Economic Research such as exists in Oxford. The fields of research must be determined in consultation with the various factors interested.

Lenders and borrowers, marketing and supply agencies, national and local agricultural institutions, the Government and the consumers, all have a vital stake in the comprehensive planning and co-ordination of agricultural credit in Israel.

An agrarian policy of a State in which more than 90% of the land is publicly owned, in which 95% of the water and power resources is administered by public bodies, in which the greater part of Government means is invested in agricultural development, in promoting an agricultural education, research and co-operation, in which an enlightened farming community with highly developed institutions exists and prospers—such a policy renders a most valuable contribution to the progress and advance of world agriculture.

INDEX